Plant Chimeras

Plant Chimeras

Richard A. E. Tilney-Bassett
Department of Genetics
School of Biological Sciences
University College of Swansea

Edward Arnold

© Richard Tilney-Bassett 1986

First published in Great Britain 1986 by
Edward Arnold (Publishers) Ltd, 41 Bedford Square, London WC1B 3DQ

Edward Arnold (Australia) Pty Ltd, 80 Waverley Road, Caulfield East,
   Victoria 3145, Australia

Edward Arnold, 3 East Read Street, Baltimore, Maryland 21202, U.S.A.

British Library Cataloguing in Publication Data

Tilney-Bassett, Richard A. E.
   Plant chimeras.
   1. Plant genetics 2. Mosaicism
   I. Title
   581.1'58   QK983

ISBN 0-7131-2936-0

Text photo set in 10/11pt Times by 🅰 Tek-Art, Croydon, Surrey.

Printed and bound in Great Britain by
Richard Clay plc, Bungay, Suffolk

Preface

In recent years exciting developments in the improvement of vegetatively propagated plants by mutation breeding, and their multiplication by tissue culture, has spawned a fresh interest in plant chimeras. The newer research is often specialized, and so the authors are unable to present the significance of their findings within the broader context of chimeras in general. Indeed, the interconnecting paths between different areas of chimeral studies, and of various types of chimera, are seldom trodden. It is hardly surprising, therefore, in the absence of suitable sources of information, to find that students of the plant sciences are often ignorant even of the existence of chimeras. In endeavouring to overcome this deficiency, and to bridge the gaps between one researcher and another, and between researcher and teacher, my prime purpose in writing this book is to increase the awareness of chimeras, and to show through the descriptions of the main types how much they have in common and yet how varied and fascinating they are. By carefully explaining the essential terms as they arise, and by keeping the use of an unfamiliar vocabulary to the minimum, the reader need have no more than an elementary knowledge of plant biology plus an interest in cultivated plants.

A book on chimeras would not be complete without a liberal use of illustrations, and I am indebted to the publishers for their sympathetic reaction to this essential need. I should like to thank the editors of Heredity for their permission to republish some illustrations from my early articles, and similarly Elsevier/North-Holland Biomedical Press, and Dr J. T. O. Kirk, for granting permission to publish a few illustrations from the second edition of The Plastids. The many new illustrations are the work of my daughter Amanda Tilney-Bassett, with advice from my wife Elisabeth, to both of whom I am very grateful. Many of the drawings are loosely adapted from published work as acknowledged in the legends. Some of the authors of the published work are now deceased, but to the others I should like to convey my sincere appreciation. Often the illustrations, adapted to my specific requirements, are samples of many more in the original paper and I urge the reader to seek out the original publications for a full appreciation of the authors' intentions. The photographs were donated by Dr F. A. L. Clowes and Dr A. Roberts, or were developed from my negatives by Mrs C. Fisher, to all of whom I am very grateful. I should also like to thank Mr J. K. Burras who permitted me to photograph examples of the very fine collection of variegated-leaf chimeras that he has maintained over many years at the Botanic Garden of the University of Oxford.

I should like to express my appreciation to the authors, and particularly

Dr F. Pohlheim, who have so kindly sent me copies of their papers, and especially to Professor Dr F. Bergann and Mrs L. Bergann who have fed my interest in chimeras for over twenty years. My thanks are also due to my research students – Paramjit, Osman, Ali, Ayad, Riadh and Baset, who have listened so patiently and responded so enthusiastically to my sporadic outbursts on the subject. Finally, I should like to thank all the family for their great forbearance during the many months of preparation.

Richard A. E. Tilney-Bassett
University College of Swansea
1986

Contents

1 The Chimera Concept

The graft-hybrid hypothesis

In the wake of the great explorers a wealth of new species was brought to Europe by plant collectors from around the world. New trees, shrubs and herbs came into our country estates, parks, botanic and private gardens (Fisher, 1982). Trading in plants became a profitable enterprise, and many nurserymen became breeders anxious to improve their stocks. This they did by the raising of new hybrids through crosses between species, and by the selection of promising new types among the progeny recombining the most favourable characters of their parents. These were skilled men, who knew their plants well, and who had a keen eye for any new sports. A sport was a spontaneous change, or mutation, in a part of the plant that created a feature not previously known within the species, their hybrids and descendants. Sports were gratuitous, valuable sources of new variation, much prized by the ambitious breeder. Charles Darwin (1868) collated an extensive list, which he classified into sports of fruit, flowers, leaves and shoots, and suckers, tubers and bulbs. At the beginning of the century de Vries (1901–03) published his famous treatise on mutation, and Cramer (1907) his account of the known cases of bud variation.

True sports are collectively quite common but individually rare, so the repetition of a seemingly identical sport was unusual. Consequently, a variation that was so frequent as to be predictable attracted attention. The 'Bizzarria' orange was such a case. This strange plant was first described in 1674 by Nati (Strasburger, 1907). Apparently, in 1644, a Florentine gardener grafted a scion of sour orange on to a seedling stock of citron. The scion did not take successfully, but a bud arising out of the callus developing on the stock grew up into a bizarre tree. On the same plant, there were leaves, flowers, and fruit, identical with the orange or with the citron, and there were compound fruit with the two kinds blended together or sectored in various ways (Penzig, 1887; Sarastano and Parrazzani, 1911). The same fruit was sometimes half orange, half citron, or three-quarters of one kind and one-quarter of the other, and when propagated by cuttings, the tree retained its peculiar character.

A second plant, that behaved in a similar way to the 'Bizzarria', originated in 1825 at Vitry, near Paris, in the garden of Adam the nurseryman. He had inserted a shield from the bark of a broom into a stock of common laburnum. The bud lay dormant for a year, and then grew with a flourish of buds and shoots, one of which grew more upright and vigorous, and with larger leaves than the usual broom; so it was selected for further

propagation, and was subsequently distributed throughout Europe as +
Laburnocytisus adamii. (The + sign is a taxonomic convention to indicate
that the plant mentioned is a graft hybrid: the × sign is a similar convention
to indicate a sexual hybrid.) The tree proved to be somewhat unstable, and
branches spontaneously reverted to both parental species in their flowers
and leaves. The tree was often described (Poiteau, 1830; Prevost, 1830;
Hénon, 1839; Braun, 1849, 1873; Caspary, 1865; Darwin, 1868; Morren,
1871; Beijerinck, 1900), but probably never better than by Darwin:

> 'To behold mingled in the same tree tufts of dingy-red, bright yellow and
> purple flowers, borne on branches having widely different leaves and manner
> of growth, is a surprising sight. The same raceme sometimes bears two kinds
> of flowers, and I have seen a single flower exactly divided into halves, one side
> being bright yellow and the other purple; so that one half of the standard-petal
> was yellow and of larger size, and the other half purple and smaller. In another
> flower the whole corolla was bright yellow, but exactly half the calyx was
> purple. In another, one of the dingy-red wing petals had a bright yellow narrow
> stripe on it; and lastly, in another flower, one of the stamens, which had become
> slightly foliaceous, was half yellow and half purple; so that the tendency to
> segregation of characters or reversion affects even single parts and organs. The
> most remarkable fact about this tree is that in its intermediate state, even when
> growing near both parent species, it is quite sterile; but when the flowers
> become pure yellow or purple they yield seed.'

When Darwin examined branches with yellow flowers they appeared to
have completely recovered their laburnum character, whereas branches
with purple flowers were not always exactly like the broom:

> 'The branches with purple flowers appear at first sight exactly to resemble
> those of *C. purpureus*; but on careful comparison I found that they differed
> from the pure species in the shoots being thicker, the leaves a little broader,
> and the flowers slightly shorter, with the corolla and calyx less brightly purple:
> the basal part of the standard petal also plainly showed a trace of the yellow
> stain. So that the flowers, at least in this instance, had not perfectly recovered
> their true character; and in accordance with this, they were not perfectly fertile,
> for many of the pods contained no seed, some produced one, and very few
> contained as many as two seeds; whilst numerous pods on a tree of the pure
> *C. purpureus* in my garden contained three, four and five seeds.'

One view was that these trees were just sexual hybrids (de Vries, 1901-03;
Strasburger, 1906, 1907) with differences in the fruits of the 'Bizzarria'
caused by vegetative segregation (Cramer, 1907). Another view was that,
as a result of grafting, the two species had actually formed a hybrid through
the union of their vegetative tissues (Braun, 1849; Caspary, 1865; Darwin,
1868). The idea, that vegetative nuclei of stock and scion could fuse to
produce a graft-hybrid with a homogeneous hybrid growing-point, and that
bud variations could arise by a somatic segregation, analogous to the
segregation of different seedling characters from sexual hybrids, was not to
be ignored. Proponents of the graft-hybrid hypothesis regarded a sexual
hybrid as unlikely because nobody had succeeded in deliberately obtaining
a hybrid between the laburnum and broom, even though both had produced
hybrids with other species, whereas they were readily grafted together. The
reversion to the parental types was also inexplicable from a sexual hybrid.

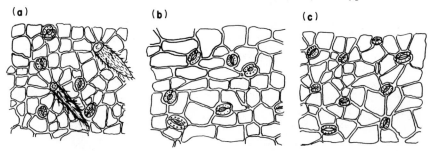

Fig. 1.1 Illustrations of the surface view of the lower epidermis from leaves of the graft hybrid + *Laburnocytisus adamii* and its constituent species. **(a)** *Laburnum anagyroides* with hairs; **(b)** + *Laburnocytisus adamii* without hairs; **(c)** *Cytisus purpureus* without hairs. (Adapted from Macfarlane, 1892.)

Additional support for the graft-hybrid hypothesis came with the discovery of a new example. A hundred-year-old tree had been discovered in a garden in Bronvaux, near Metz, in which a medlar scion had been grafted on to a hawthorn stock. Two branches had developed from the graft union between stock and scion and these were both intermediate in character, although not identical (Jouin, 1899; Koehne, 1901; Bornmüller, 1932; Guillaumin, 1949). One branch, called + *Crataegomespilus asnieresii*, more nearly resembled hawthorn, and the other, called + *Crataegomespilus dardari*, more nearly resembled medlar. The correlation between grafting and the development of shoots with an intermediate type of growth seemed to confirm the graft-hybrid hypothesis, but increasingly evidence was accumulating against it.

Macfarlane (1892) carefully examined the anatomical structure of + *L. adamii* and its parents and found the comparison between the epidermis of broom and the graft-hybrid of particular interest:

> 'But the very striking resemblance which the epidermis of the hybrid portion has to that of *C. purpureus*, not only in the general structure of the cells, but in the size and structure of the cell nucleus, the distribution of the stomata, and specially of hairs (Fig. 1.1), would seem at first sight to prove that the hybrid portion was wrapped round, so to speak, by an epidermis of *C. purpureus*.'

Fuchs (1898), Laubert (1901) and Noll (1907) extended the anatomical investigations without doubting the graft-hybrid interpretation either, yet adding to the evidence that made quite a different explanation possible.

The general habit of + *L. adamii* was of a small tree more like the common laburnum than the bushy broom, and the leaves and flowers were closer to laburnum than to broom. The reversion to pure broom was less common than to laburnum (Beijerinck, 1901; Laubert, 1901), wounding enhanced the frequency of laburnum shoots, and when resting buds were induced to sprout, they frequently developed into laburnum (Beijerinck, 1908). The graft-hybrid had fairly well formed pollen grains but the embryo sac usually degenerated (Tischler, 1903), even so, rare seeds were obtained and these germinated solely into laburnum seedlings (Noll, 1907; Hildebrand, 1908). It therefore began to appear that + *L. adamii* was not so much a hybrid as a mixture predominantly laburnum in character but with an epidermis typically broom. Still further doubt was raised by Strasburger (1907, 1909), who examined the nuclei of + *L. adamii* and its two parents and found that all three had the same chromosome number, $2n = 48$; if cell fusion had taken place cells of the hybrid would have been expected to have double the chromosome complement. With little support for the sexual hybrid hypothesis, and decreasing confidence in the graft-hybrid hypothesis for + *L. adamii*, the time was right for a better alternative, yet the breakthrough came from a different quarter.

The chimera hypothesis

An experimental approach to testing the graft-hybrid hypothesis was begun by Winkler (1907), who made saddle grafts between the tomato and the black nightshade, using each species as stock or scion. After union, the scion was removed by a transverse cut at the junction of stock and scion. A callus soon grew across the exposed surfaces out of which adventitious buds arose. Some buds developed into shoots resembling tomato or nightshade, but one graft resulted in a plant in which the shoot was divided longitudinally into two halves – one half was composed of tomato and the other of nightshade. Recalling the Chimaera of Greek mythology, which was a fire-breathing monster, the foreparts of whose body were those of a lion, the middle parts those of a goat, and the hind parts those of a dragon, Winkler called his new plant, which was composed of two genetically distinct tissues, a chimera (Fig. 1.2).

Five other grafts between tomato and nightshade developed shoots with intermediate structures, which Winkler initially believed to be the sought after graft-hybrids and to which he gave specific names (Winkler, 1908, 1909). It did not seem to have struck him as odd that fusion between identical nuclei could develop into several hybrids of varying structure, but he was soon to modify his view.

Confirmation of Winkler's chimera concept came shortly afterwards with quite unrelated experiments on white-margined, variegated-leaf, zonal pelargoniums (Fig. 1.3). After crossing green and variegated plants, Baur (1909a) obtained a mixture of green, white, and variegated progeny. Some of the variegated seedlings were divided longitudinally such that one half of the axis had green leaves and the other white leaves, and leaves which were half green, half white were occasionally produced on the border between the two. These plants were like Winkler's tomato-nightshade chimera and Baur concluded that:

> 'The plants have, therefore, quite evidently a sectorially divided growing-point just as in the well known chimeras of Winkler.'

These plants Baur called sectorial chimeras.

Baur also made sections through the leaves of white-margined pelargoniums and found that a layer of colourless palisade or spongy mesophyll cells formed an unbroken skin completely enclosing the inner core of green cells. No colourless skin occurred in green plants. When he

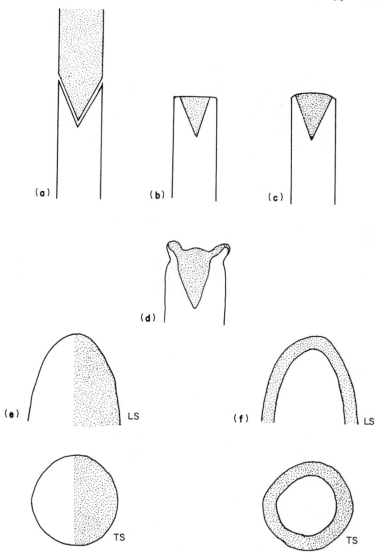

Fig. 1.2 Diagrams illustrating stages in the formation of sectorial and periclinal shoots after grafting a scion of one species onto a stock of another. **(a)** A nightshade scion (shaded) is grafted onto a tomato stock (unshaded). **(b)** After union, a transverse cut exposes a small wedge of scion tissue embedded in the stock. **(c)** Cell division leads to the growth of callus across the exposed surface of stock and scion. **(d)** Adventitious bud development in the callus shows the nightshade growing alongside the tomato (left) and over the tomato (right). **(e)** The left hand bud has developed into a sectorial chimera with nightshade and tomato tissue growing alongside each other, as seen in longitudinal section (**LS**) and transverse section (**TS**). **(f)** The right hand bud has developed into a periclinal chimera with a skin of nightshade enveloping a core of tomato, as seen in longitudinal section (**LS**) and transverse section (**TS**).

Fig. 1.3 Two examples of variegated-leaf, zonal pelargonium seedlings. In both plants there is differentiation of the leaves into a margin of one colour and a centre of another. The upper plant has leaves with a green margin and white centre. The lower plant exhibits a reciprocal arrangement with a white margin and green centre. (From Tilney-Bassett (1963b), with the permission of the editors of *Heredity*.)

selfed his plants, green cultivars and occasional green shoots arising on the variegated cultivars gave green seedlings. Similarly, white-margined cultivars and occasional white shoots arising on these gave white seedlings. It thus became clear to Baur that the colour of the seedlings corresponded with the colour of the tissue forming the skin, and so the white skin and the green core of the white-margined plants must occupy distinct layers of the shoot. This relationship was only altered when the white skin replaced the green core, or the green core displaced the white skin, to form wholly white or green shoots respectively. Normally, Baur argued, the white layer must be continuous from the leaves back into the apical growing-point and into the flowers where it formed the germ layer (Fig. 1.4). Moreover, the same structural relationship of the white and green tissues could be obtained whenever variegated seedlings developed shoots with white-margined leaves just like the parental cultivars; these confirmed Baur's second conclusion:

'With that is indeed the nature of white-margined plants clear, they are likewise chimeras, not chimeras with sectorially divided growing-point but chimeras with periclinally divided growing-point, in short they may be called Periclinal Chimeras.'

The significance of Baur's hypothesis was immediately recognized because it agreed so well with the findings of the late nineteenth century anatomists. They had found that the cells at the tip of higher plant shoots were frequently organized into two or more layers, which were

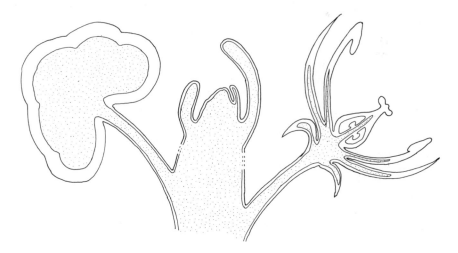

Fig. 1.4 Diagram to illustrate Baur's concept of the periclinal chimera. The leaf margin, on the left, traces back to a skin within the layered growing-point, which includes both epidermal and subepidermal layers, and is continuous with the margins and tips of the floral parts, including the germ tissues. The skin completely encloses the underlying core. In a variegated-leaf chimera, the leaf margin is white and the central tissues green. But there is no colour differentiation between skin and core in either the growing-point or the floral parts as the genes determining the contrasts in pigmentation are not expressed in these organs.

superimposed one above the other like a series of cones from which the body structure of the plant was derived. Thus Baur was able to explain the structure of white-margined leaves by assuming that the pattern found in the leaf was a lateral and flattened extension of the layered structure already present at the growing-point (Fig. 1.4).

After Winkler had demonstrated that graft-hybrids between tomato and nightshade could be obtained experimentally, Strasburger (1909) rejected his former belief that the graft-hybrids were really sexual hybrids. But he could not accept the simplicity of Baur's periclinal theory. Instead, he assumed that there had been a more intimate and irregular arrangement of the tissues from stock and scion to produce a hyperchimera, comparing it more to a lichen than to a seed plant. Baur (1909a, 1909b, 1910, 1911) was not put off, and promptly suggested that all Winkler's tomato-nightshade grafts were periclinal chimeras too. By comparing four of these with each other and with their parental species, Baur was able to work out their probable chimeral structure. For each plant, the character was determined for the epidermis by matching the surface hairs to those of the appropriate parent, for the germ layer by the seedlings obtained after self-pollination, and for the innermost tissue by the shapes of the leaves (Table 1.1). Winkler (1910a) accepted Baur's interpretation for these four plants, yet still retained his belief in the concept of graft-hybrids.

Baur (1910) also suggested that + *L. adamii* was a periclinal chimera with an epidermis of broom over a core of laburnum. Similarly, after the comparison between the graft-hybrids of + *Crataegomespilus* and their parents by Noll (1905), Baur was able to interpret these as periclinal chimeras too – with a skin of medlar over a core of hawthorn. The anatomical investigations of Buder (1910, 1911) soon led to confirmation of the periclinal structure for + *L. adamii*. Buder was able to trace the general distribution of broom and laburnum tissue in young and old shoots, petioles and leaves by precipitating tannins with potassium bichromate. As broom is rich in tannins and laburnum has none, this made an excellent marker, which clearly revealed the periclinal structure. Moreover, it showed up patches where the epidermis from broom had thickened, or a sector of laburnum had thrust through the skin. The reversions to broom originated in periclinal divisions within the young epidermal cells of the shoot apex, which normally grew by anticlinal divisions. The more frequent reversions to laburnum probably arose because division of these cells always replaced the epidermis whenever it was damaged.

Two further, and very elegant, demonstrations of the periclinal structure were found in the development of the cork and the petals. The formation of cork in young branches arose in the epidermis in broom, below the epidermis in laburnum, and in both positions in + *L. adamii*. The petals were pigmented in the epidermal cells by purple anthocyanins in broom, by yellow plastids in laburnum, and by purple anthocyanins in + *L. adamii*. In addition, laburnum and the graft-hybrid, but not broom, contained dark brown cells forming the honey guides two cells deep below the epidermis. These observations, together with the earlier records of Macfarlane, were fully consistent with the periclinal interpretation of the graft-hybrid, which

Table 1.1 The four distinct graft chimeras between tomato and nightshade as synthesized by Winkler, their origin, three characteristic features of each, and the interpretation of their periclinal chimera structure by Baur.

Name	Origin		Features			Structure		
	Stock	Scion	Leaf surface	Seedlings	Leaf form	Epidermis	Germ layer	Core
Solanum tubingense	S. nigrum	L. esculentum	esc	nig	nig	esc	nig	nig
Solanum proteus	S. nigrum	L. esculentum	esc	esc	nig/esc	esc	esc	nig
Solanum koelreuterianum	L. esculentum	S. nigrum	nig	none	esc	nig	esc	esc
Solanum gaertnerianum	L. esculentum	S. nigrum	nig	nig	esc/nig	nig	nig	esc

was therefore not a hybrid at all, but a graft chimera or species chimera.

The acceptance that + *L. adamii* and four of Winkler's experimental graft-hybrids were actually periclinal chimeras did not lead to a complete abandonment of the graft-hybrid hypothesis. Daniel (1904, 1909, 1914, 1915) attributed several cases to this origin including the still inadequately resolved + *Pyrocydonia* types, derived from grafts of pear and quince, and the + *Amygdalopersica* types, derived from grafts of almond and peach (Daniel and Delpon, 1913). These cases were not unambiguous. Daniel was supported by Weiss (1930) in regarding + *Pyrocydonia* as a graft-hybrid, even though Weiss accepted the chimeral explanation for several other graft-hybrids, nor was Swingle (1927) opposed to the theoretical possibility, although he regarded proof as lacking; Krenke (1933) was more sceptical. After further experimental grafting in the Solanaceae family, Winkler (1934, 1935, 1938) was convinced that he had obtained true graft-hybrids. Cramer (1954) seemed to concur but Brabec (1949, 1954) declared that Winkler's observations could still be explained by the occurrence of chromosome abnormalities arising naturally, or by chromosome loss or gain during callus formation. Fortunately, the pursuit of these elusive hybrids, which may now be achieved in some species by the quite different technique of protoplast fusion, did not stop progress in the understanding of graft chimeras.

Graft chimeras

Following the general acceptance of the chimera concept, interest in graft chimeras continued, partly owing to the search for the elusive graft-hybrid, and partly to investigate further properties of individual cases (Table 1.2). As detailed discussions are covered in reviews (Swingle, 1927; Weiss, 1930; Rudloff, 1931; Krenke, 1933; Neilson-Jones, 1934, 1937, 1969; Guillaumin, 1949; Cramer, 1954; Brabec, 1965) I shall keep my remarks brief.

A fresh look at 'Bizzarria' convinced Tanaka (1927a, 1927b) that the plant was a periclinal chimera with an epidermal skin of sour orange over a core of citron. In his opinion the earlier emphasis on its more bizarre features has been misleading; reversions to orange or citron, and breaches of the orange skin by the underlying citron tissue (Fig. 1.5), certainly occurred, but not so often as to completely break down its essential periclinical structure.

A further confirmation of the epidermal species of + *L. adamii* came from a biochemical observation by Keeble and Armstrong (1912). They showed that the epidermal cells of the petals of + *L. adamii* and of broom contained an oxydase enzyme that responded directly with benzidine to produce a colour reaction, whereas there was no corresponding response with the epidermal cells of laburnum unless hydrogen peroxide was added. A similar test in the petal veins showed that the tissue beneath the epidermis of + *L. adamii* was of laburnum origin. Together with the many anatomical observations, there could be no doubt that the graft chimera had a skin of broom over a core of laburnum.

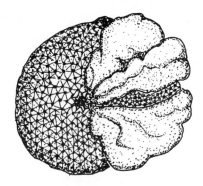

Fig. 1.5 Two views of fruit from the 'Bizzarria' orange. **Left** A section through the periclinal chimera showing the sour orange in the centre, derived from the inner epidermis, and the pure citron pulp and rind towards the outside, derived from the core. **Right** Surface view showing a sector with an orange outer epidermis on one side, breached by citron on the other. (Adapted from Tanaka, 1927a.)

Darwin's observation that branches with purple flowers were not always like the pure species, and had flowers in which the basal part of the standard petal showed a trace of yellow, suggests that these branches had both an epidermal and a sub-epidermal layer of broom over a much reduced core of laburnum. This means that the vegetative and flowering shoots of the graft chimera are developed from three layers. Normally, there is a single layer of *Cytisus purpureus* and two inner layers of *Laburnum anagyroides*. So the structure of the periclinal chimera is usefully defined, from outside to inside, as *Cytisus Laburnum Laburnum*, which is conveniently abbreviated *CLL*.

The bud variation described by Darwin can then be explained on the assumption that duplication of the outer layer by periclinal divisions results in the replacement of the laburnum tissue beneath, in other words there is a change in the periclinal structure from *CLL* to *CCL*. Similarly, reversions to pure broom can be assumed to arise by a continuation of this process through further periclinal divisions and replacement bringing about the change from *CCL* to *CCC*. Conversely, reversions to laburnum can be explained as arising through the displacement of the outer layer of broom by the inner layers of laburnum, which is represented as the change from *CLL* to *LLL*.

A continuing interest in + *L. adamii* has been maintained by Sneath's (1968) numerical taxonomic study, by an analysis of the distribution of osmiophilic bodies in the inner and outer epidermis of the gynoecium (Boeke and van Vliet, 1979), and by Bergann's (1952) discovery of a new type. According to the investigations of Bauermeister (1969) and Pohlheim (1969a), there appears to have been a spontaneous chromosome doubling in the broom skin to give a chromosome number of $2n = 96$, and correlated increases in cell size, nuclear size, and the number of chloroplasts per cell. This created a new periclinal chimera, which they called 'Laburnum B', from which a new type of broom of higher ploidy has been isolated, which they called 'Neocytisus'.

Doubt as to the graft-hybrid nature of the two + *Crataegomespilii* was raised by Meyer (1915), who found that both they and their parents had 2n = 34 chromosomes. Moreover, differences in the chromosome morphology of the two species led him to recognize the chimeral structure of the graft-hybrids. Even so, Weiss (1925) and Haberlandt (1926, 1927) began to doubt the chimeral explanation owing to a discrepancy between the shapes of the leaf epidermal cells in surface view and those of the presumed medlar parent. Cramer (1954) later suggested that they had failed to appreciate the extent to which the shapes of the epidermal cells were in a juvenile state, or were modified by the underlying tissue of another species. More telling evidence for their chimeral structure came from the few pure hawthorn seedlings derived from + *C. asnieresii* (Baur, 1930; Haberlandt, 1930), but not from + *C. dardari* which set no seed.

Probably the most useful study was the thorough investigation of leaf development by Lange (1933), in which he identified tissues by the size of cells and nuclei, and by the use of a differential colouring method. This *showed that + C. asnieresii* had a single layered skin of *Mespilus germanica* over a two layered core of *Crataegus monogyna – MCC*, while + *C. dardari* had a two layered skin of *Mespilus* over a single layered core of *Crataegus – MMC*. These interpretations agreed with the more medlar-like appearance of the latter and the more hawthorn-like appearance of the former, and with the observed frequencies of bud variations, + *C. asnieresii*, with a single layered skin, rarely reverted to medlar – *MCC to MMM*, whereas + *C. dardari*, with a double layered skin, often did – *MMC to MMM*.

Meyer (1915) also observed a branch of + *C. asnieresii* growing out of + *C. dardari – MMC to MCC*, but not the reverse. This was curious as the reverse change had occurred in + *L. adamii*, and reversion to both parents was frequent (Lange, 1933), and Haberlandt (1931) had even observed fruits which had reverted to the pure hawthorn tissue of the core. The further analysis of these plants was continued by Haberlandt (1934a, 1934b, 1935, 1941) and Bergann (1951, 1956). Pohlheim and Pohlheim (1974) have drawn attention to a yellowing pattern that became visible in the summer leaves of + *C. dardari* corresponding to the area of mesophyll occupied by the inner core of hawthorn, and which appeared similar to a variegated-leaf pattern in the pelargonium 'Kleiner Liebling'. In a fresh look at the + *Crataegomespilii*, Byatt *et al.* (1977) argue that the influence of the chimeral partners on each other has often been underestimated.

Other cases of graft chimeras between *Crataegus* and *Mespilus*, having an independent origin, have been described by Daniel (1914), Seeliger (1926) and Hjelmqvist (1937). Bergann and Bergann (1984b) have described two new graft chimeras, which they themselves grew, and which they have called + *Crataegomespilus potsdamiensis* 'Monekto' and 'Diekto', with a thin or thick skin respectively, of medlar over a core of hawthorn. They differ from the classical graft chimeras because instead of using the common hawthorn with single white flowers, they grafted the variant 'Pauli', which has double

Table 1.2 List of natural graft chimeras, their composition, probable structure, and their main investigators.

Name	Composition and structure		Investigators
	Skin[1]	**Core**	
+ *Amygdalopersica* (types)	*Prunus dulcis/P. persica*		Daniel and Delpon, 1913
Camellia + 'Daisy Eagleson'	*Camellia sasanqua* L I	*C. japonica*	Pohlheim, 1976; Stewart et al., 1972
Citrus + 'Bizzaria'	*Citrus aurantium* L I	*C. medica*	Cramer, 1907; Penzig, 1887; Sarastano and Parrazzani, 1911; Strasburger, 1906, 1907; Tanaka, 1927a, 1927b; de Vries, 1901–03
Citrus + 'Kobayashi-mikan'	*Citrus unshiu* L I	*C. natsudaidai*	Yamashita, 1979, 1983
+ *Crataegomespilus asnieresii*	*Mespilus germanica* L I	*Crataegus monogyna*	Bergann, 1951, 1956; Baur 1930; Bond 1936;
+ *Crataegomespilus dardari*	*Mespilus germanica* L I/II	*Crataegus monogyna*	Bornmüller, 1932; Byatt et al., 1977; Daniel, 1909, 1914; Fischer, 1912; Haberlandt, 1926, 1927, 1930, 1931, 1934a, 1934b, 1935, 1941; Hjelmqvist, 1937; Jouin, 1899; Koehne, 1901; Lange, 1933; Maurizio 1927; Meyer, 1915; Noll, 1905; Pohlheim and Pohlheim, 1974; Sahli, 1916; Seeliger, 1926; Weiss, 1925
+ *Crataegomespilus potsdamiensis*:			
'Monketo'	*Mespilus germanica* L I	*Crataegus monogyna*	Bergann and Bergann, 1984b
'Diekto'	*Mespilus germanica* L I/II	'Pauli'	
+ *Laburnocytisus adamii*	*Cytisus purpureus* L I	*Laburnum anagyroides*	Baur, 1910; Beijerinck, 1900, 1901, 1908; Buder, 1910, 1911; Boeke and van Vliet, 1979; Braun, 1849, 1873; Caspary, 1865; Darwin, 1868; Fuchs, 1898; Hénon, 1839; Hildebrand, 1908; Keeble and Armstrong, 1912; Laubert, 1901; Macfarlane, 1892; Morren, 1871; Noll, 1907; Poiteau, 1830; Prevost, 1830; Sneath, 1968; Strasburger, 1907; 1909; Tischler, 1903
+ ''Laburnum B'	*Cytisus purpureus*-$4x$[2] L I	*Laburnum anagyroides*	Bergann, 1952; Bauermeister, 1969; Pohlheim, 1969
—	*Populus bachofenii* L I	*P. nigra* 'Italia'	Kosichenko and Petrov, 1975
—	*Populus nigral/P. suaveolens*		Vasil'ev, 1965
—	*Populus suaveolens/P × canadensis* 'Regenerata'		Vasil'ev 1965
+ *Pyrocydonia* (types)	*Pyrus communis/Cydonia oblonga*		Daniel, 1904, 1915; Weiss, 1930
+ *Syringa correlata*	*Syringa vulgaris* L I	*S. chinensis*	Hjelmqvist, 1947

[1] When known, the graft chimeras are classified according to whether the skin is derived from the outermost layer, or tunica, of the apical meristem (L I), or from the two outermost layers, or tunicas (L I/II). For a full explanation see Chapter 4.

[2] The skin of this broom has cells with four times the basic chromosome complement (4x) instead of the usual two copies of each chromosome (2x).

red flowers, and which regularly reappears as a vivid sporting branch.

An interesting study reported by Bond (1936) was of the effect of the two chimeras in resisting infection to fungi to which medlar was immune and hawthorn susceptible. To what extent did a single layered, or double layered skin of medlar protect the hawthorn core? Fischer (1912) reported that + *C. asnieresii* was susceptible to *Gymnosporangium confusum*, which Sahli (1916) confirmed, whereas + *C. dardari* was less readily infected. This showed that the resistant epidermis alone was not an effective barrier, but with two layers fungal penetration was greatly reduced.

The method did not appear to have validity for testing periclinal chimeras as experiments by Maurizio (1927) with *Podosphaera oxycanthae* did not separate the two chimeras on the basis of a differential resistance to infection. Klebahn (1918) had been rather more successful in his attempt to infect Winkler's tomato-nightshade chimeras with *Septoria lycopersicii* and *Cladosporium fulvum*, to which tomato is susceptible and nightshade resistant. The graft chimeras + *S. tubingense* and *S. gaertnerianum* were both immune, whereas + *S. koelreuterianum* and + *S. proteus* were both killed. When one has regard to the structure of these chimeras (Table 1.3), it is apparent that the subepidermal layer is the key; a subepidermal layer of nightshade resisted infection, whereas a subepidermal layer of tomato was susceptible.

Further confirmation of the periclinal structure of Winkler's tomato-nightshade graft chimeras came from a detailed study of leaf development, and generally from chromosome analysis (Lange 1927). Tomato has $2n = 24$ and nightshade $2n = 72$ chromosomes; hence a true graft-hybrid was expected to have, after fusion of vegetative cells, $2n = 96$ chromosomes. An analysis of the pollen grain nuclei confirmed the species suspected of forming the subepidermal layer in each, and there was no evidence of any hybrid layer.

Another graft chimera, + *S. darwinianum*, was more difficult to interpret; Winkler thought it was a hybrid as cells were found with $2n = 48$ chromosomes, which he suggested might have arisen as a somatic reduction from $2n = 96$. Neilson-Jones (1934, 1969) considered it more likely that $2n = 48$ had arisen by a somatic doubling of the tomato complement, and pointed out that Winkler (1916) had himself found evidence of doubling after callus formation. Hence + *S. darwinianum* was quite possibly a graft chimera with two layers of nightshade over a core of tetraploid tomato.

In addition to Winkler's chimeras, graft chimeras were made by several other workers utilizing a variety of species (Table 1.3). Some of these were none too stable. Heuer (1910) found that the graft chimera between tomato and aubergine was continually reverting to the tomato core, and Lieske (1921, 1927) had the same experience with the chimera between tomato and *S. dulcamara*. Jörgensen and Crane (1927) found that in some of their chimeras the leaves were buckled and malformed. This was a particular feature in combinations in which the outer component formed a skin from a species with small simple leaves, while the core was from a species bearing large compound leaves (Fig. 1.6). Krenke (1933) found the chimera between tomato and *S. memphiticum* rather unstable and found leaves of either pure type, or sectors, or changes in the proportions of the

Table 1.3 List of synthetic graft chimeras within and between the genera *Lycopersicon*, *Nicotiana*, *Saracha* and *Solanum* of the family Solanaceae, their probable structures, and their main investigators; in chronological order of synthesis.

Name of species	Structure			Investigators
	Epidermis	Germ layer	Core	
S. darwinianum	nig	nig	esc-4x	Fucik, 1960; Günther, 1954, 1957a, 1957b, 1962; Juncker and Mayer, 1974; Klebahn 1918; Krüger, 1932; Lange, 1927; Mayer-Alberti, 1924; Mayer et al., 1973; Winkler, 1908, 1909, 1910a, 1910b
S. gaertnerianum	nig	nig	esc	
S. koelreuterianum	nig	esc	esc	
S. proteus	esc	esc	nig	
S. tubingense	esc	nig	nig	
L. esculentum-S. dulcamara	esc	dul	dul	Heuer, 1910
S. melongena-L. esculentum	mel	esc	esc	
S. dulcamara-L. esculentum	dul	esc	esc	Lieske, 1921, 1927
S. tuberosum-L. esculentum	tub	esc	esc	Jörgensen, 1927
S. nigrum 'Gracile'-S. sisymbrifolium	gra	gra	sis	Jörgensen and Crane, 1927
S. nigrum 'Guineense'-L. esculentum	gui	esc	esc	
S. nigrum-L. esculentum	gui	gui	esc	
	nig	esc	esc	
	nig	nig	esc	
L. esculentum-S. nigrum	esc	nig	nig	
Saracha umbellata-L. esculentum	umb	esc	esc	Krenke, 1929
S. memphiticum-L. esculentum	mem	esc	esc	Krenke, 1933
	mem	mem	esc	
L. pimpinellifolium-S. nigrum	pim	nig-4x	nig-4x	Brabec, 1960
L. esculentum-L. peruvianum	esc	per	per	Günther, 1961
S. pennelli-L. esculentum	pen	esc	esc	Clayberg, 1975; Heichel and Anagnostakis, 1978
N. tabacum-N. glauca	tab	gla	gla	Marcotrigiano and Gouin, 1984b
	tab	tab	gla	

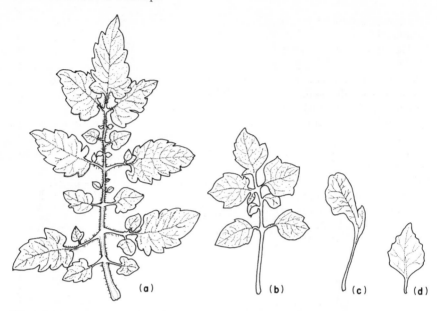

Fig. 1.6 Illustrations of typical leaves from two pure species plus two graft chimeras between them to show the effect of a thin or thick skin of one species over a core of the other. **(a)** Pure *Lycopersicon esculentum*; **(b)** The thin skinned chimera – *luteum, esculentum, esculentum*; **(c)** The thick skinned chimera – *luteum, luteum, esculentum*; **(d)** Pure *Solanum luteum*. (Adapted from Jörgensen and Crane, 1927.)

components. He also found that a skin of *S. memphiticum* provided frost protection to the sensitive tomato core. Cytochimeras with changes in ploidy were also found occasionally.

An important use of these synthetic chimeras was to study the origin of tissues in different parts of the plant. Mayer-Alberti (1924) found that in + *S. proteus* the subepidermal layer of tomato tissue was responsible for much of the mesophyll of the leaf blade, while the deeper layer of nightshade tissue was responsible for the main vascular bundles in the petiole and leaf blade. Lange (1927), who extended anatomical studies to + *S. koelreuterianum* and + *S. gaertnerianum*, found some differences between the chimeras in respect of their contributions but, in general, he found that the inner core tissue did not run into the leaf margins, which were therefore derived largely from the sub-epidermal layer, but it did contribute to parts of the mesophyll, as well as to parts of the main and lateral veins in the centre of the leaf. Krüger (1932) found that all three layers participated in the growth of fruit. The subepidermal layer was decisive for the final form of the fruit, while the innermost layer was important in controlling the size. The normal development of one species was influenced by the juxtaposition of another. For instance, the normal hairiness of the tomato epidermis was reduced if immediately beneath there was a subepidermis of cells derived from the glabrous nightshade. Jörgensen and Crane (1927) and Krenke (1933) both found that the pigmentation of the fruit was modified by the properties of the cells underlying the cells of manufacture.

Besides periclinal chimeras with a single or double skin layer, Krenke (1933) believed that there might, on rare occasions, even be a three layered skin. He emphasized that one, two or three layers implied that this was the thickness of the skin at the growing-point. Although in vegetative and flowering shoots, or in fruit, the skin was not normally thinner than at the growing-point, it could be and often was very much thicker owing to the occurrence of periclinal divisions during development. Moreover, there was not necessarily correspondence between the thickness of the skin in any organ and that at the growing-point, except that generally, but by no means always, the outer layer at the growing-point was mainly responsible for the epidermis alone. Inner layers tended to be more variable and, particularly in fruit, Krenke found that the tissues of the two components were greatly intermixed to form an irregular periclinal layered structure rather than an orderly one. Sometimes, he believed, the skin in the growing-point was so thick that the lateral organs developed from the skin alone, and consequently the chimeral nature of the stem was not suspected from the general appearance.

In recent years synthetic graft chimeras have been used to study problems of incompatibility (Günther, 1961), and to demonstrate the decisive role played by the epidermis in the perception of and response to light (Mayer *et al.*, 1973; Junker and Mayer, 1974), and to assess the stomatal response to light (Heichel and Anagnostakis, 1978). In quite another direction, Clayberg (1975) found that a skin of insect resistant *S. pennelli* protected the underlying tomato core against the greenhouse white fly, *Trialeuroides vaporariorum*, to which tomato is highly susceptible; but a similar resistance to the potato aphid, *Macrosiphum euphorbiae*, was not achieved.

Graft chimeras of other species have been reported on several occasions (Table 1.2). Hjelmqvist (1947) has suggested that an old garden form of lilac, + *Syringa correlata*, had an epidermis of *S. vulgaris* over a core of *S. chinensis*. Vasil'ev (1965) reported on two poplar graft chimeras. One with a skin of *Populus nigra* over a core of *P. suaveolens*, and the second with a skin of *P. suaveolens* over a core of *P. × canadensis* 'Regenerata'. A third example had an epidermis of *P. bachofenii* over a core of *P. nigra* 'Italia' (Kosichenko and Petrov, 1975). Stewart *et al.* (1972) described a graft chimera, *Camellia* + 'Daisy Eagleson', which had an epidermis of hexaploid *C. sasanqua* 'Maiden's Blush' over a core of *C. japonica*; Pohlheim (1976) has discussed the occurrence of petaloid styles in this chimera. Finally, Yamashita (1979, 1983) used scanning electronmicroscopy and the analysis of esterase and peroxidase isozyme patterns to show that the *Citrus* + 'Kobayashi-mikan' had an epidermis of *C. unshiu* over a core of *C. natsudaidai*. In some of these cases the epidermis was derived from the scion and in others from the rootstock.

An interesting development has been the experimental synthesis of tobacco chimeras. This was achieved by Marcotrigiano and Gouin (1984a) through mixing cells derived from callus cultures homozygous for the wild-type, *Su/Su*, and the semi-dominant sulphur mutant, *su/su* (Tilney-Bassett,

1984). The contrast between the green and white cells of the two genotypes enabled the four chimeras to be readily picked out from the 1317 non-chimeral regenerating shoots. On the other hand, the two investigators (Marcotrigiano and Gouin, 1984b) failed to obtain any chimeras by mixing callus of *Nicotiana tabacum* and *N. glauca*, whereas they did manage to derive three mericlinal graft chimeras between these species after using *N. glauca* as the stock. In two cases, by selective pruning, they were able to isolate and propagate axillary periclinal shoots, one with a thin tobacco skin, *TGG* and one with a thick tobacco skin, *TTG*.

The first chapter has concentrated on graft chimeras because these played an important role in establishing the chimera concept, but before looking at other types of chimera it will be useful to establish some basic ideas on the relationship between the nature of chimeras and that of the meristems that makes their existence possible.

2 Meristems

The apical meristem

Coniferous and flowering plants consist of a root, generally growing below ground, and a shoot, generally growing above ground. Roots are not prominent in chimera studies, except when they give rise to adventitious shoots, or as chromosomal mixochimeras, and so they will not be considered further. The shoot, which has a stem that is usually upright, bears two kinds of lateral organ – the leaves and buds. The leaves, which occur in a wide variety of shapes, are characterized by their determinate growth and their dorsi-ventral symmetry. The buds are characterized by their potentially indeterminate growth and their radial symmetry. The leaves develop as outgrowths from the stem at positions called nodes, and the buds usually develop singly in the axils of the leaves. During vegetative growth buds remain either dormant, or develop into new shoots. During reproductive growth buds develop into cones or into inflorescences and flowers.

The growth of vegetative and flowering shoots is brought about by the frequent divisions of many cells localized in areas called meristems. The primary meristem is the apical meristem found at the tip, or apex, or apical growing-point, of the shoot. Further elongation of the stem is achieved by division of cells of the sub-apical region, or by a true intercalary meristem, while increase in girth is brought about by two lateral meristems – the vascular cambium and the phellogen – which give rise to secondary tissues – the wood and the bark. In some contexts the apical meristem refers specifically to the summit of the tip above the youngest leaf primordium – the protomeristem or metrameristem – which includes the leaf initials and their most recent derivatives (Esau, 1965); in other contexts the term is used in a more general way to include the whole region of the shoot tip in which cell division is active (Figs 2.1 and 2.2).

The apical meristem becomes established in the developing embryo. Its size and shape is extremely variable between species. It is commonly a paraboloid, but it may also be flat, slightly concave, conical, or much elongated. At the level of the youngest leaf primordium, the diameter varies from 3 mm in *Cycas revoluta* down to a fiftieth of this size. During development, within a single species, there may be as much as a 7-fold variation in width and a 3-fold variation in height, and corresponding changes in volume. Some of these changes in the continually shifting population of meristematic cells are attributable to the stage of development, especially of the transition from vegetative to flowering shoot, and some to cyclical changes related to the periodic formation of leaf

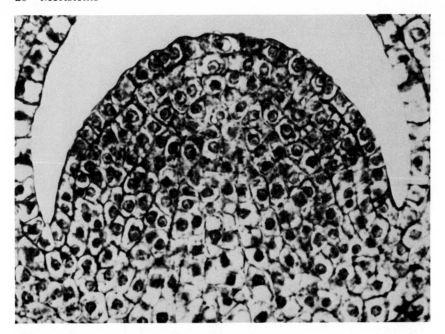

Fig. 2.1 Longitudinal section through the paraboloid shoot apical meristem of *Hypericum calycinum*. At the summit three cell layers are visible above the non-layered corpus, but it is not possible to decide from a single section whether or not this is a one-, two-, or three-layered apex. On the left hand side there is evidence that the inner layers are dividing periclinally, possibly at the start of the next primordium. (Photograph kindly supplied by Dr F.A.L. Clowes.)

primordia. At certain seasons, in many woody plants, the apical meristem and sub-apical region cease growing for a period and the tip forms a terminal bud and becomes dormant. In other species, the whole tip aborts during unfavourable conditions, and dormant terminal buds are not formed.

Following Wolff's (1759) recognition of the tip as the zone from which the rest of the shoot grows, the idea developed that all cells are descendants of a limited number of initial cells. An initial is a cell that divides into two daughter cells, one of which remains in the fixed apical initial zone while the other, being in a relatively peripheral position, is added to the meristematic tissue that eventually differentiates further. The simplest organization is to have one initial, and when a single apical cell was found in many vascular cryptogams, such as the ferns, it was thought that a single initial of this kind would be found in higher plants too (Fig. 2.3a). This has remained an attractive idea to many investigators, but it is not supported by the finding of any specific cell distinguishable from among a number of similar cells at the summit.

An alternative, the histogen theory, was developed by Hanstein (1868). He proposed that initials, which were arranged one above another at the summit, gave rise to an inner core of irregularly arranged cells covered by a variable number of regular, mantle-like layers. He further proposed that this meristem consisted of three parts, the histogens, which were

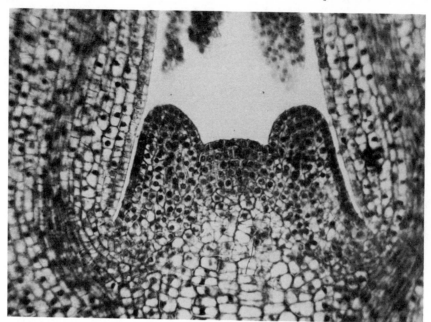

Fig. 2.2 Longitudinal section through the flat shoot apical meristem of *Ligustrum ovalifolium*. The layering of the apex is sharp, and cells at the flanks of the meristem are clearly contributing to the development of the leaf primordia. (Photograph kindly supplied by Dr F.A.L. Clowes.)

differentiated by their origin and course of development. The outermost layer, the dermatogen, formed the epidermis; the middle part, the periblem, composed of the rest of the mantle layers, formed the cortex; the innermost part, the plerome, formed the vascular tissue and pith (Fig. 2.3b). Unfortunately for the proposal, it turned out that the relationship between the kinds of differentiated tissue in the mature shoot and the apical layers from which they were descended was by no means so precise. On the contrary, graft and cytochimeral studies in particular, have helped to show that the way in which a cell differentiates is not dependent upon its direct lineage back to a specific initial within the meristem, but solely by the position it finds itself within the differentiating tissues.

The third theory of apical organization was developed by Schmidt (1924) and his teacher Buder (1928). They divided the apical meristem of flowering plants into two zones: The outer zone, the tunica, consisted of one or more peripheral layers of cells in which divisions were predominantly anticlinal – at right angles to the surface; the inner zone, the corpus, consisted of a central core of cells in which divisions were anticlinal and periclinal – parallel to the surface – and in other planes too (Fig. 2.3c). The anticlinal divisions of the tunica allowed for surface growth, and the variable divisions of the corpus allowed for growth in volume. Each layer of the tunica arose from a small group of separate initials, and the corpus had its own initials

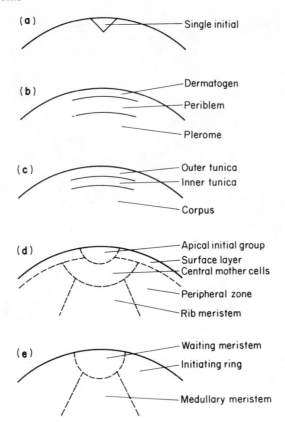

Fig. 2.3 Diagrams illustrating various theories on the location of initials within shoot apices of higher plants. **(a)** Single apical initial. **(b)** Initials within histogens. **(c)** Initials within tunica layers and corpus. **(d)** Initials within zones. **(e)** Initials surrounding inert waiting meristem.

beneath those of the tunicas. Although superficially alike, the tunica-corpus theory differs from the histogen theory in that is does not imply any direct relationship between the organization of the cells at the apex and subsequent tissue development. In practice, problems have often arisen with regard to the actual number of tunica layers for particular species, and hence for the line of demarcation between tunica and corpus. Even so, the theory has proved very useful for our description and understanding of chimeras.

A new structure arises with respect to the apical meristem of gymnosperms, as few species have a layer that can be interpreted as a tunica. In most species the outermost layer has cells dividing both anticlinally and periclinally and so it is not distinct from the inner cells to which it gives rise. This led Foster (1938) to the recognition of a zonation based on the cytological and histological appearance of groups of cells and on their degree of meristematic activity. At the extreme tip he recognized an apical initial group of cells from which all other zones were ultimately derived. The zone immediately below these initials constituted the central mother cells which stained less densely, were rather vacuolated, and were believed to divide infrequently. The centrally situated derivatives of these mother cells led into a rib meristem, which gave rise to the pith, while the more lateral derivatives formed a densely staining and highly meristematic peripheral zone encircling all the inner zones (Fig. 2.3d). In longitudinal sections the peripheral zone appears, and is frequently referred to, as a flank meristem. A similar zonation has been observed in many angiosperms (Clowes 1961).

Another movement away from the layered concept for the apical meristem in flowering plants was developed in the French school of Buvat (1952, 1955). He believed that the distal zone of the apex was inert during vegetative growth and only became active at the change to reproductive growth; he therefore called it the waiting meristem (*méristème d'attente*). The real growth of the vegetative shoot was assumed to come from an initiating ring (*anneau initial*) of meristematic cells in the peripheral zone where the leaf primordia arose, and from the medullary or pith meristem (*méristème medullaire*) (Fig. 2.3e). A disadvantage of the model is that, while emphasizing the zones of greatest meristematic activity, it distracts from the importance of the true initials within the waiting meristem as the ultimate source of all other cells of the shoot. Moreover, the model does not stress the layering of these initials, which does occur, if not in all plants, at least in those in which stable chimeras are found.

In order to reduce the confusion brought about by the proliferation of terms used to describe shoot apices, Gifford and Corson (1971) have contructed a useful diagram of a hypothetical dicotyledonous shoot apex labelled to illustrate the more important concepts and synonyms (Fig. 2.4).

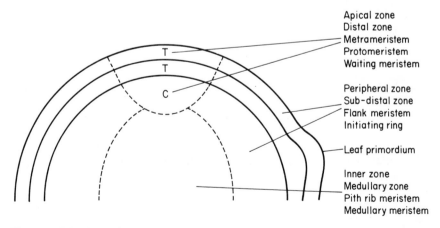

Fig. 2.4 A longitudinal section through the shoot apex of a typical dicotyledon indicating many of the descriptive terms commonly used. Mitotic activities are not indicated. (Adapted from Gifford and Corson, 1971.)

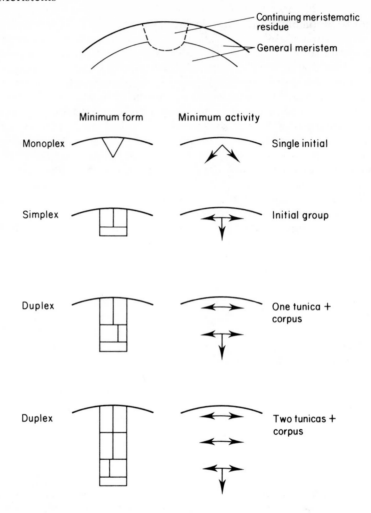

Fig. 2.5 A longitudinal section of a shoot apex within higher plants defined in terms of a continuing meristematic residue and a general meristem. The structure and behaviour of the three types of apex are illustrated on the left by minimum form and on the right by minimum activity. (Adapted from Gifford and Corson, 1971.)

The zonation of the apex is reflected by the separation of an apical zone from a subjacent peripheral zone and pith-rib meristem. The layering of the apex is expressed by the division of the apical zone into a tunica and corpus.

As the theories of apical organization evolved, it became clear that not all apices were alike. In order to classify the types of apex, Newman (1961, 1965) first made the distinction between a continuing meristematic residue – the initials – and the general meristem (Fig. 2.5). When an initial divided, one of the two daughter cells continued to function as an initial within the continuing meristematic residue, while the other daughter cell became part of a general meristem. Emergence of new cells from the continuing meristematic residue, the source of cellular structure, was slow, but continuous and of long duration, whereas in the general meristem, the region of elaboration, cells passed through a phase of rapid division which was continuous, but of only short duration. Based on the form of the continuing meristematic residue, apices were classified into the following three types (Cutter, 1971; Gifford and Corson, 1971).

1. *The monoplex apex*, characteristic of ferns; it has initials in the superficial layer only. Any one cell division contributes to growth in length and breadth, and only one initial is needed.

2. *The simplex apex*, common in gymnosperms; it has a zone of initials in the superficial layer only. Anticlinal divisions are required for surface growth and periclinal divisions for growth in length and bulk.

3. *The duplex apex*, common in angiosperms; it has two superimposed zones of growth – the tunica and corpus – each taking its origin from initials for each zone. Anticlinal divisions are the rule in the one or more tunica layers, while divisions in various planes occur in the corpus.

In order to form stable chimeras, the subject of this book, the species involved must have either simplex apices in which periclinal divisions are rather infrequent as in a few gymnosperms, or duplex apices with one or more tunicas as in many monocots and dicots. In some plants, longitudinal sections through shoot apices suggest the presence of more than two tunicas. Nevertheless, owing to the increasing frequency of periclinal divisions in the third and successive tunicas over extended periods of development, there may be insufficient stability to maintain the multi-layered structure, and so the innermost tunicas become regarded as derivatives of the outer tunicas and so part of the corpus. Moreover most chimeras are classified, not by a direct examination of the apical meristem, but by the number of distinguishable tunicas plus corpus that contribute to the appearance of the leaves, flowers or other organs. Hence the innermost layer that actually contributes to the lateral organ is regarded as derived from the corpus, and successive outer layers as derived from the one, two or rarely three tunicas. What generally counts is therefore the number of initial layers in the apical meristem to which growth of the lateral organs trace back, rather than the number of initial layers to which the growth of the stem traces back; these are not bound to agree.

Leaf, bud and floral meristems

In the case of most gymnosperms, the leaves trace back to mixed divisions in a small group of cells in the surface layer of the flank meristem, while in a few the leaves trace back to mixed divisions in both surface and subsurface layers. A few angiosperms are like gymnosperms, but in many the leaves are initiated by periclinal divisions in a small group of cells in the subsurface layer of the flank meristem in positions according to the system of leaf arrangement; these divisions are followed by similar divisions in the

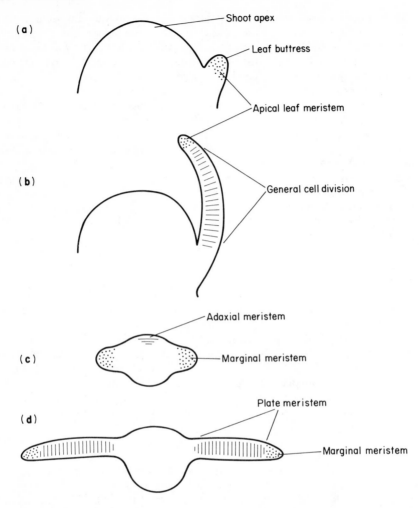

Fig. 2.6 Diagram illustrating the location of meristems within the developing leaf. **(a)** Longitudinal section of an early leaf buttress showing growth by the apical leaf meristem. **(b)** Longitudinal section of a later leaf buttress with intercalary growth by general cell division. **(c)** Transverse section showing lateral outgrowth of a young leaf by the marginal meristem, and growth in thickness of the midrib and petiole by the adaxial meristem. **(d)** Transverse section of the young leaf showing further lateral growth by the plate meristem. (Adapted from Cutter, 1971.)

third layer and by anticlinal divisions in the surface layer. With usually from one to three layers forming the leaf, and with only anticlinal, or with mixed, divisions in the outer layer, a range of leaf structure is possible, which is reflected in the range of chimeras found. The divisions initiating a leaf primordium produce a leaf buttress in the side of the shoot apex from which the leaf grows upwards and outwards and finally expands laterally. The upward growth proceeds by extensive cell division, initially in apical regions as the leaf buttress elongates (Fig. 2.6a), and later through general cell division throughout the primordium in dicots (Fig 2.6b), or by a well-defined intercalary meristem at the base of the primordium in most monocots. After the dorsiventral leaf primordium has reached a certain height, lateral outgrowths, which constitute the marginal meristem, appear on either side (Fig.2.6c); the activity of these establishes the number of layers of cells in the lamina. The cells laid down by the activity of the marginal meristem divide anticlinally and expand the lamina laterally. As each cell divides to give a small plate of cells, the whole lamina functions as a plate meristem (Fig. 2.6d). In many dicots a strip of cells below the surface on the side forming the stem – the adaxial meristem – divides periclinally and contributes to the thickness of the leaf in the region of the petiole and midrib (Fig. 2.6c). The wide variety of leaf form is attributable to the relative activity and duration of these meristems functioning either simultaneously or sequentially.

At the end of cell division, expansion and differentiation, a typical leaf appears as a lateral organ attached to the stem by a petiole that continues out into the central midrib and veins of the vascular system, which supports and supplies an extensive, flattened lamina. In transverse sections, the surface of the leaf consists of an upper, or adaxial, and a lower, or abaxial, epidermis, either or both of which, according to species, are punctuated by pores. The opening and closing of these pores, or stomata, are controlled by specialized epidermal cells – the guard cells. The normal epidermal cells usually contain a few, small, widely dispersed chloroplasts, whereas in the guard cells the much larger chloroplasts fill the cells. Beneath the upper epidermis, a region of thin walled cells elongated in the transverse plane and containing numerous chloroplasts forms the palisade tissue (Fig. 2.7).

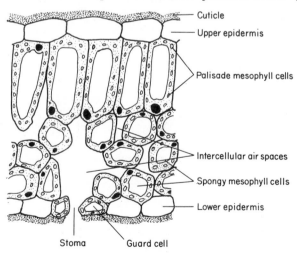

Fig. 2.7 A transverse section through a leaf showing the main features of its cellular structure.

Beneath the palisade and above the lower epidermis, a region of irregularly-shaped cells with sparser chloroplasts and many intercellular spaces forms the spongy mesophyll.

In most angiosperms, lateral bud primordia usually develop in the axils of the leaves, slightly later than the subtending leaf primordia. Bud initiation involves a combination of anticlinal divisions, in one or more of the superficial layers of the shoot axis, and of periclinal and other divisions, in the third or deeper layers. As buds frequently arise in deeper layers than the leaves, the innermost cells of the bud may trace back to slightly deeper initials in the corpus of the apical meristem than do those of the leaves. This coordinated growth in surface area and bulk pushes the bud out between its leaf and the stem. As the axillary bud develops into a shoot, its apical meristem is gradually organized, normally repeating the pattern of the parental shoot apex with its leaves, in turn, tracing back to the same tunica-corpus initials.

Some buds arise on stems, roots, and leaves, particularly on cuttings, in places having no direct connection with the apical meristem. These adventitious buds often arise entirely from superficial tissues, or entirely from deep tissues, and so do not reflect the two or three layered structure of the shoot apex; ultimately they trace back to only one of the apical layers. Their occurrence has often proved useful for the analysis of chimeras, and as a means of producing a pure shoot, free from the chimeral structure of its parental plant.

The inflorescence is frequently regarded as a modified shoot in which the axis is homologous with the stem and the lateral organs with the leaves; there are many parallels between the two. The reproductive apex of higher plants results from a more or less extensive reorganization of the vegetative apex, and growth of the inflorescence proceeds through the activation of a similar range of meristems. As with vegetative shoots, the variation in the appearance of inflorescences is vast (Cutter, 1971; Esau, 1965). Of particular significance is that inflorescence apices retain the layering of their vegetative apices so there is an important consistency between the two regions (Fig. 1.4).

This brief account of the organization of meristems has provided the basic theory for an understanding of chimeral structures, which we shall now examine. We shall begin by looking at the origin of chimeras, namely the transient sectorial and mericlinal chimeras, and what we can learn from them. Afterwards, the remainder of the book is largely concerned with the more stable and permanent periclinal chimeras.

3 Sectorial and Mericlinal Chimeras

Origin and appearance

Observant plantsmen often see examples of plants in which a spontaneous mutation has produced a white sector, or streak, owing to a blockage in the formation of the normally green chlorophyll pigments. Usually the mutation is fleeting and limited to the leaf in which it occurs. Sometimes the mutation occurs in the apical meristem; it then develops as a white sector, or streak, running down the side of the shoot and embracing any leaves and buds, or parts thereof, that have arisen from primordia developing within its pathway. When the mutation occurs in a species with a monoplex or simplex apex, or early in embryo development, the mutant lineage forms a solid white sector in the side of the shoot, and the plant is a sectorial chimera (Fig. 3.1;2). When the mutation occurs in a species with a duplex apex, the mutant lineage is usually confined to the tunica layer, or the corpus, in which the mutated cell arose. As the sector is not solid the

Fig. 3.1 Kinds of chimera shown diagrammatically in transverse section (inner circle) and longitudinal sections (outer ring) through the apical meristem of a typical higher plant. **(1)** Solid for first genotype. **(2)** Sectorial chimera. **(3)** Solid for second genotype. **(4)** Periclinal chimera. **(5)** Mericlinal chimera. (From Kirk and Tilney-Bassett, 1978.)

plant is not a truly sectorial chimera; the mutant lineage is inside or outside a normal layer, or between normal layers, and the sector is therefore periclinal in relation to other layers, and the plant is more accurately referred to as a mericlinal chimera (Fig. 3.1;5). Indeed, the majority of so-called sectorial chimeras are probably mericlinal (Jörgensen and Crane, 1927; van Harten, 1978).

The size of the mutant sector depends upon where the mutation arises. When the mutation occurs in one of a small number of initials at the summit, the angle of the sector is large. Thus from a mutation in one of two initials we expect a mutant lineage encompassing half the circumference of the stem, from one of three initials a third, from one of four initials a quarter, and correspondingly smaller fractions with even more initials. With increasing distance from the summit up to the initiation of leaf primordia the apex broadens, and so the more peripheral the mutation the narrower the streak.

The width of the mutant sector helps to determine the appearance of the chimera. A wide sector regularly embraces whole leaves, and parts of leaves, whereas a narrow streak merely touches the occasional leaf. In plants in which the leaves develop one above another, white sectors cast almost identical patterns on successive leaves, but when the leaves follow a spiral sequence the shafts of white cut the leaves in varying places in relation to the shifting origin of the leaf primordia. Bain and Dermen (1944) used colchicine solution to induce tetraploidy in the apical meristem of runners and uprights in the 'Searls' cranberry. Some of the treated shoots became diploid-tetraploid mericlinal chimeras. When the epidermis was affected, they were able to distinguish, by visual inspection of the lower surface at low magnification, those leaves and parts of leaves that were diploid or tetraploid. Hence they were able to score the position of each leaf along the stem and record its ploidy. This led them to the discovery that the ploidy pattern of the leaves was repeated after each thirteenth leaf. In other words, if leaf 1 had a wholly tetraploid epidermis, so had leaves 14, 27 and 40. The position of the leaves around the stem followed a spiral, and as five complete turns were required for each renewal of the pattern, the distribution of leaves on the stem followed a 5/13 phyllotaxy.

In a similar manner, the appearance of mutant leaves, and parts of leaves, in other plants is expected to depend on the angle subtended by the mutant sector, the length of the arc of the leaf initials around the circumference of the stem, and the specific phyllotactic pattern. The most frequent fractions belong to the Fibonacci series 1/2, 1/3, 2/5, 3/8, 5/13, 8/21 etc., in which each value of the nominator and denominator is a sum of the two values that precede it. Although characteristic for each species, the phyllotaxy is not always constant and may vary between shoots or change during growth.

The mutant lineage is not always so sharp as pictured. Sometimes the meristem expands laterally and broadens the mutant sector, or conversely the normal cells expand laterally and narrow the mutant sector. When the two populations of normal and mutant cells invade each other extensively, especially from adjacent layers, the white sectors become divided into tongues between the green so increasing the complexity of the patterns.

Indeed, the shapes of white sectors within the laterally expanded, flattened leaf blades are excellent markers of the patterns of growth within the meristems as they reveal the major cell lineages in a perfect, visual display (Dulieu, 1968, 1969; Stewart and Dermen, 1975).

The longevity of a sector is related to its place of origin. A mutant sector descending from an apical initial lasts a long time unless, exceptionally, the initial is displaced by a neighbouring normal cell. By contrast, a sector descending from a mutant cell well down the flank meristem is short lived owing to its quick replacement by the growth of the normal meristematic cells above.

Each mutation arising in the shoot tip creates a new sectorial or, more probably, mericlinal chimera, and there are reports of a wide range of these. Some mutations have led to sectorials in, for example, flower colour (Cook, 1930) or root colour and shape (Lamprecht and Svensson, 1949) in the carrot, or for yellow and red periderm in the bark of the white fir (Zobel, 1974). Most recorded cases are in the leaves, flowers and fruit of ornamental flowers and shrubs, and in orchard trees, some of which are described in later chapters. In a few plants, which will be discussed after the next section, observations of the fraction of the sector within the circumference of the shoot has proved a useful guide as to the probable number of initials for that tissue.

Mutation induction

The induction of mutations in higher plants is widely practiced. The physical mutagens used include X-rays, gamma rays, and fast and thermal neutrons, while the chemical mutagens include ethyl methane sulphonate (EMS), diethyl sulphate (DES), ethylene imine (EI), propane sultone, N-methyl-N-nitroso urethane (MNU) and some of its related compounds, and also sodium azide (Gottschalk and Wolff, 1983). Many other compounds have mutagenic activity, but most workers prefer to use well tried and tested ones. The preferred technique for sexually propagated crops is to irradiate dry seeds prior to germination as these are easier to handle than moistened seeds, or shoots, and chemicals are more inclined to produce unwanted side effects, especially loss of vigour and reduced fertility. Nevertheless, once the right conditions for a plant are worked out, chemical mutagens are extremely useful and widely used. About 99% of all mutants are due to recessive mutations, and mutant lineages are not normally discernible in the M_1 generation. It is then necessary to self or intercross the M_1 plants in order to obtain, in the M_2 generation, segregation from the heterozygous lineages of the M_1 generation. Here a serious difficulty arises. In most treated seeds any induced mutations, if they occur at all, affect only one initial cell within the embryonic growing-point; independent mutations within a single seed occur in different initials. Hence, depending on the number of initials, most mutations produce lineages of limited size and only a segment of the flowers and flowering shoots becomes heterozygous. Each affected plant is a hidden mericlinal chimera and so, as the M_2 segregation occurs among seed collected from the invisible mutant segment, the seed must be harvested without selection. It follows that the homozygous recessive mutant progeny are likely to occur in a much less favourable ratio than the classical 3 normal:1 mutant expected from selfing a pure heterozygote, and so the best strategy for obtaining the mutants needs to be carefully considered (Gottschalk and Wolff, 1983).

In the remaining 1% of cases the mutation is dominant and is expressed in the M_1 generation. Alternatively, many ornamentals are heterozygous for genes controlling important characters like flower colour. Consequently, the mutation of a dominant allele of a heterozygote creates a homozygous recessive lineage which is expressed directly. In such cases the mutagen creates an exposed mericlinal chimera in the M_1 generation from which the mutant lineage is more easily isolated, or from which a stable periclinal chimera may be selected. The induction of chimeras in the M_1 generation of heterozygous plants has often been used to test the mutagenic properties of chemicals, rather than to investigate the properties of chimeras, although sometimes periclinal chimeras have developed (Cornu, 1970). Where available, the induction of mutation in haploids, as in *Thuja plicata* 'Gracilis' (Pohlheim, 1977a) or the zonal pelargoniuim 'Kleiner Liebling' (Pohlheim *et al.*, 1972) also has the advantage of an immediate response in the M_1 generation (Pohlheim, 1981b).

With the spread of techniques for growing plant material in tissue culture (Dodds and Roberts, 1982) treatment of cell callus and suspended cell populations is likely to gain in importance as the starting point for mutant induction. *In vitro* regeneration of shoots from the callus of hypocotyl and cotyledonary explants of tomato seedlings heterozygous for the semi-dominant xanthophyllic-2 mutant, gave rise to about 10–15% chimeras (Seeni and Gnanam, 1981). During callus growth the allelic combination *Xa-2/xa-2* was prone to somatic segregation, presumably owing to irregular chromosome distribution, gene conversion, or somatic recombination, giving rise to albino, *Xa-2/Xa-2*, and green, *xa-2/xa-2*, lineages in addition to the original yellow-green. A similar result was obtained with stem plants cultured from a heterozygous tobacco (Seeni and Gnanam, 1976). No chimeras occurred in shoots derived from the callus of homozygous plants. Other examples, and the possible causes for the instability of these heterozygotes are discussed more fully by Kirk and Tilney-Bassett (1978). Leaf and flower variegations also appeared to be associated with chromosome instability in regenerants of some somatic hybrids obtained by protoplast fusion, as for example between *Nicotiana knightiana* and an albino mutant of *N. tabacum* (Maliga *et al.*, 1978).

The mutagen treatments discussed so far are aimed at altering nuclear genes; on other occasions the plastid genes are the main targets. Many of the variegated-leaf chimeras have arisen as spontaneous plastid mutants followed by sorting-out into periclinal chimeras. Experimentally, plastid mutation, leading to chimera formation, has been chemically induced by EMS, particularly in tobacco (Dulieu, 1965, 1967a, 1967b, 1970; Deshayes, 1973), and more recently in a number of crop species – peas, carrots, soybeans, lentils, and radishes (Miller *et al.*, 1980, 1984). Another effective compound is nitrosomethylurea (NMU), which has been successfully applied in snapdragon, sunflower and the zonal pelargonium (Kirk and Tilney-Bassett, 1978), and more recently in *Lycopersicon esculentum*, *L.hirsutum* and *L. pennelli* (Hosticka and Hanson, 1984), and *Saintpaulia ionantha* (Pohlheim, 1974, 1979, 1980b, 1981a; Pohlheim and Beger, 1974; Pohlheim and Pohlheim, 1976). Since the mutagens that induce plastid gene mutations may also induce nuclear gene mutations, a careful analysis of the mutant is needed before definite conclusions can be drawn as to the exact mutational site (Kirk and Tilney-Bassett, 1978; Tilney-Bassett, 1984).

Apical initials

At the shoot apex the tunicas cover the corpus as thin skins over a solid cone. The initials at the summit of a tunica have to divide, by anticlinal divisions, in such a way as to produce a succession of cells descending down and all around the broadening circumference. How many initials are required at the summit?

When Moh (1961) irradiated coffee seeds with up to 2.8×10^{13} cm^{-2} of thermal neutrons or 16 krad gamma rays, he obtained a high frequency of morphological mutants in the M_1 generation; these were mostly 'angustifolia' (*ag*) mutants characteristically producing dwarf growth and small or elongated leaves. Surprisingly, rather than appearing as sectors, the whole plant, with the possible exception of the epidermis, became mutant, irrespective of the radiation dose used. It therefore seemed as if a single cell in the corpus acted as the primary initial for the early stages of shoot development. Broertjes and Keen (1980) developed a stochastic model to describe the process of apex formation in adventitious buds, from which they could estimate the expected relative percentages of chimeras. In practice, the actual frequencies of chimeras in various crops, either propagated *in vivo* or *in vitro*, was lower than expected. This led them to postulate that either the apices of adventitious shoots were formed from only a single epidermal cell of the meristem or callus, or that apex formation was not a random process but a process in which a particular cell was 'chosen' – perhaps the first dividing cell – to assume a leading position within the meristem, probably surrounded by its genetically identical daughter cells. It is possible that the induction of mutations in coffee, after irradiation, followed a somewhat similar pattern of behaviour. Supposing the irradiation treatment inhibited the coordinated division of the apical initials, they might then have been replaced by the first cell among them, or close by, that sprung into division, and if that cell had undergone a mutation, the whole growing-point would have become mutant. In effect, the growing-point would have behaved as if derived from a single initial, but this need not reflect the situation as found in the growing-point of a non-irradiated, non-adventitious shoot, where more than one initial might be the norm.

Darwin (1868) described an apple in which one half was red with an acid taste and a peculiar odour, while the other half was greenish-yellow and very sweet. Another apple was half green, half brown skinned, and Clayberg (1963) described an apple half red, half yellow. Other examples of a similar division are Winkler's half tomato, half nightshade graft sectorial chimera, and a half triploid, half hexaploid sectorial in *Solanum × juzepczukii* (Howard, 1961a). Two potato tubers have been described, one a half white-splashed pink, half full colour for the outer tunica of 'Gladstone' (Howard, 1961b), the other a half pink, half purple 'Arran Victory' (Howard, 1969a). A mutation in the 'Blazer' tomato produced a half red, half tangerine pericarp – the tissue below the epidermis which includes the seeds and flesh. Genetic analysis of the seeds from the two sectors showed that the red tissue was heterozygous, Tt, for a dominant nuclear gene and the tangerine sector was homozygous recessive. The colour of the leaves indicated that the mutation was probably to a virescent-tangerine mutant – Tt to $t^{vir}t$ (Walkof, 1964). All these instances suggest that the two halves of the various apices are derived from separate lineages and possibly from two initial cells.

A one third black russet sectorial in the 'Ruby' orange (Robinson, 1927), a one third sectorial in the 'Deglet Noor' date palm that affected the trunk, foliage, inflorescence and fruit (Mason, 1930), and sectors of one third in the potato (Klopfer, 1965b), are all indicative of the possibility of three initials. The number of initials in the subepidermal layer of 'Bonus' barley was estimated by Lindgren et al. (1970) after 5 to 8 krad γ-irradiation of dry seeds, or after treating them with 0.12 to 0.25% EMS solution. They were able to examine the M_1 generation directly by scoring plants for waxy mutants. Normal haploid pollen grains, Wx, contain amylose and amylopectin and are stained blue to black in a dilute iodine solution, whereas pollen grains carrying the recessive waxy allele, wx, lack amylose and are therefore stained a reddish brown. As the two types are readily distinguished, the size and frequency of sectors can be determined by scoring the anthers of flowering spikes for waxy mutants. After making various corrections, and assumptions, the mean number of initial cells was equal to the inverted value of the mean sector size, which gave values after different treatments between 2.3 and 3.6, that is around three initials. The distribution of sector sizes estimated from the scoring of the segregation ratios for chloroplast mutants was in accordance with the waxy mutant data and supported the estimate of three initials.

Verkerk (1971) irradiated seed of the tomato 'Moneymaker' with 3 to 4 krad fast neutrons at a dose rate of 1 krad per hour. The plants of the normal-looking M_1 generation were topped above the second truss to encourage the development of side shoots, which were removed, rooted, and planted out. Seed was collected from trusses on both mother and separated side shoot plants, and all carefully numbered for their topographical position. The cotyledons and first two to three leaves of the M_2 generation were scored for morphological and colour mutations. Analysis of the M_2 generation showed that the majority of mutations occurred in the region of the first five leaves of the M_1 generation and were singles, which arose below the apical meristem from cells in the flank meristem or leaf primordia. Thereafter, there was a transition through leaves six to ten to sectors, in which the same mutation was picked up more than once, indicating that different fruit, trusses, or side shoots shared a common lineage owing to the mutation having arisen in an apical initial. In the region of, or above, the first truss on the mother plant, at about the tenth leaf, Verkerk scored mostly one or two mutant sectors and occasionally three mutant sectors, which led him to suggest that there was probably no more than three initials in the subepidermal layer of the embryo apical meristem. Unfortunately, in his rather brief account of this interesting approach to the initial problem, Verkerk did not make it clear how many non-mutant sectors grew alongside the mutant ones, and hence whether or not the three mutant sectors embraced the complete circumference of the growing-point. Without further information, it is questionable that the maximum of three mutant sectors actually reflected the low probability of three or more initials of one apex mutating independently at the same time, rather than indicating that there were no more than three of them.

Strong evidence for the number of initials in a tunica layer came from observations of *Epilobium hirsutum*, which showed that adjacent half leaves were more often similarly variegated than the two halves of the same leaf. Michaelis (1957) argued that the two halves of the leaf, either side of the midrib, came from different cell lineages, whereas the adjacent halves of two adjacent leaves, oriented at right angles to each other because of the decussate leaf arrangement, had a common origin (Fig. 3.2a). This arrangement was easily accounted for by assuming that leaf development traced back to a group of four initials, and this was fully supported by the apparent organization of cells as seen in anatomical sections (Bartels, 1956, 1960a, 1960b, 1966).

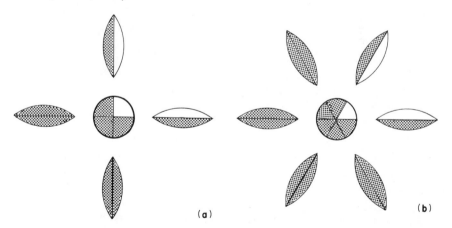

(a)   (b)

Fig. 3.2 Two models showing the relationship between the genotype of cells within a group of apical initials and the development of the leaves. **(a)** Four initials and a three quarters green, one quarter white sectorial. **(b)** Six initials and a five sixths green, one sixth white sectorial. In the left model, typical of *Epilobium hirsutum*, the decussate leaf arrangement prevents leaves forming between those shown, but in the right model, typical of potato and possibly privet, leaves develop all round the circumference owing to the spiral phyllotaxy. Consequently, a pure white leaf can be formed from the single white initial lineage in **(b)** but not in **(a)**. No significance is attached to the shape of the initial cells.

A sectorial for a one half entire leaf, one half pinnate leaf in the vegetative shoot of the 'Majestic' potato led Howard (1966b) to assume that the leaves traced back to two initials in the inner tunica. By contrast, in a sectorial for yellow and green leaves, Howard (1971b) observed sectors of about one sixth, which included leaves wholly of one colour – green or yellow – as well as individual half and half leaves; hence he felt obliged to assume a minimum of six initials (Fig. 3.2b). The idea that, besides two or three initials, the tunicas of the potato shoot might sometimes have six initials seems improbable and contrary to any concept of the fixity and stability of the initial group. A more likely hypothesis is to assume that the sectorials were misinterpreted, and that the fractions represented the ways in which a constant number of initials were split. Thus six initials can accommodate a half and half sectorial (3:3 split), a one third two thirds sectorial (2:4 split), and a one sixth five sixths sectorial (1:5 split) without any change in the size of the initial group (Fig. 3.2b). Support for six initials can also be drawn from the observations of Stewart and Dermen (1970a). They noticed that in a number of variegated plants, including the californian and chinese privets, change in the size of sectors that extended for many nodes was often from 1/2:1/2 to 1/3:2/3. They believed that this represented a change in the number of initials from two to three, that is a group of two initials became reorganized as a group of three initials. An alternative explanation is that the change actually represents a difference of one sixth (one half minus one third), which could be achieved by the alteration of one of six initials without any change in their number. Support for up to six initials is found in the shapes of cells.

Cells within a tissue appear to the eye as a pattern of linked polygons constructed according to the following basic laws (Dormer, 1980).

The surface is divided into polygons such that
    (i)   Two sides meet at a wall
    (ii) Three cells and three walls meet at a corner
The space is divided into polyhedrons such that
    (iii) Two faces meet at a wall
    (iv) Three cells and three walls meet at an edge
    (v) Four cells, six walls, and four edges meet at a vertex

The number of sides per polygon and the number of faces per polyhedron depends upon such factors as the tissue geometry, the size of the cells, and whether counted just before or after cell division. Allowing for this variation, it is possible to determine that the average number of sides per polygon is exactly six, while the average polyhedron is an orthic tetrakaidecahedron with fourteen faces in which a regular hexagon is surrounded by hexagons and squares alternately (Fig. 3.3). Cells at the surface of a tissue are below average 11-hedrons, equivalent to the equal division of a 14-hedron, but still with an average of six sides per face. In the limiting case in which small surface cells overlie a single enormous internal cell, the surface cells reduce towards the minimal attainable 8-hedrons. Bearing in mind these anatomical constraints, an impression of the initial cells within a tunica is derived by viewing the tissue from above, or in

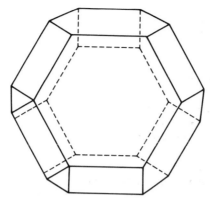

Fig. 3.3 Diagram of an orthic tetrakaidecahedron (Dormer, 1980), or truncated octahedron (Wenninger, 1971), with fourteen faces in which a regular hexagon is surrounded by hexagons and squares alternately.

transverse section, as one or more interlocking hexagons (Fig. 3.4). A group of two initials appears unsatisfactory. The symmetry is bilateral, rather than radial, and so does not accommodate the full circumference of the apex; there is only a single common wall and no central corner. By contrast, three initials form a perfect, radially symmetrical arrangement fulfilling all the criteria of any three cells within a pattern of linked polygons; it is a highly stable grouping in which each cell has equal contact with its partners. A group of four initials has structural stability, but imperfect radial symmetry; it would appear to suit the decussate leaf organization. A group of five initials can only be constructed approximately by a ring of five hexagons surrounding a central pentagon. This arrangement is probably not very stable as it is geometrically unsound. As the angles at the corners of pentagons and hexagons are 108° and 120° respectively, at each of the five corners where two hexagons and the pentagon meet the total angle is 348° and not the full 360° required. Hence five initials appears unlikely. In order to maintain the hexagonal rule, and have genetic stability, the next group of initials is a ring of six (Fig. 3.4d). It is immediately apparent that the ring of six, like the ring of five, surrounds a seventh, or first, initial in the centre. This places us in a quandary, how should we regard this organization? In theory, we can argue that the central hexagon represents a single initial from which all six surrounding cells are ultimately derived. In practice, division of the central initial has to be rather infrequent or it would not be possible to develop a moderately stable mericlinal chimera for which two lineages derived from at least two initials are required. In my opinion we should not be too dogmatic as to our definition of an initial, but rather we should consider that in this situation we have to compromise. There are aspects of the behaviour of chimeras that are best explained on the assumption of a ring of six initials, and aspects in which a role for the central initial can be invoked. Returning to the examples of plants, like the potatoes, with sectors of a half, a third, and a sixth, these can easily be explained on the basis of the varying ratios of cells of two different genotypes among a group of six

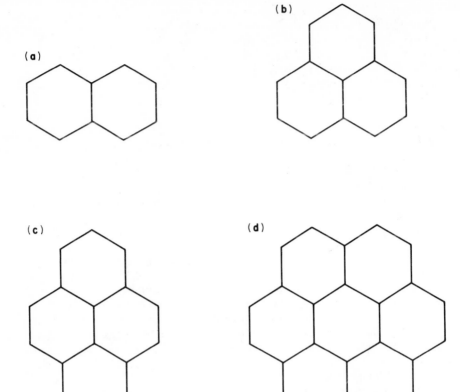

Fig. 3.4 Models of groups of initials with hexagonal-shaped cells as seen in transverse section through a tunica layer. Divisions normally cut off new cells anticlinally to the outer walls of the initials. **(a)** Two initials. **(b)** Three initials. **(c)** Four initials. **(d)** Six initials surrounding a central cell which functions as an initial rarely.

initials. So long as the central cell does not divide the sectors remain stable. But supposing, as in the variegated privets, three initials are genetically green and three genetically white, making a half and half sectorial, and supposing also that the central cell is genetically green. Then if this central cell occasionally divided, anticlinically, it would produce a green daughter cell that would be internal to and so bound to replace a cell from the ring of six. On half the occasions the replaced cell would be one of the white initials and consequently the ring of six would be changed from three green three white to four green two white. In other words, we can explain the varying sizes of the sectors by the ratio prevailing in the ring of six initials, and the rare, sporadic changes in the ratio by the occasional participation of the central cell.

A ring of six hexagonal initials is completely surrounded by an outer ring of twelve cells. This may be significant, as it is the number required to provide one initial for each half leaf of the 'Majestic' potato. Sectorial chimeras in the potato showed that each half leaf, either side of the midrib, required a separate initial, and as a leaf appeared to occupy one sixth of the circumference, the complete circle would require twelve cells. It thus appears that the development of a whole leaf, or two half leaves, traces back either to a single cell, or to a pair of adjacent cells, within the inner ring of six initials, and individual half leaves to one of the outer twelve cells. Hence if five of the six initials are genetically green and one genetically white, a leaf with both halves derived from the white initial is wholly white, but a leaf with one half derived from the white initial and the other half derived from an adjacent green initial is a white-green sectorial. By the same argument, as a group of three initials is completely surrounded by an outer ring of nine hexagonal cells, a single leaf requiring two of these cells, one for each side of the midrib, might be expected to occupy two ninths of the circumference. Steffensen (1968) induced white sectors in maize plants, which have a single tunica and corpus, by irradiating seed of the heterozygotes y_1/pastel-8549 and wd/yellow-green (Yg_2) with 15 or 30 krad γ-rays. He then carried out a detailed sector analysis of the homozygous recessive white tissue patterns induced in many of the surviving, mature plants. This led him to the conclusion that each leaf primordium had an arc of about two thirds the circumference of the stem, and that a total of seven to nine cells were required to encircle the apical meristem. A group of four initials is encircled by an outer ring of ten hexagonal cells, but the photograph of four cells in *Epilobium hirsutum* (Bartels, 1956) departs from the model. Instead, each cell appears five rather than six sided, and the four initials are surrounded by an outer ring of eight rather than ten cells. The mismatch appears to be accommodated by a degree of flexibility and stretching in the cells which conform only approximately to the same shape.

One difficulty is to comprehend how a spontaneous mutation often affects several initials at the same time. Often the answer is that the initials are not altered by a new mutation at all, but by an instability in the layer structure of an existing chimera. Periclinal divisions in the cells of an unseen mutant layer in which the phenotype is not expressed, or is not visible, may easily replace or displace cells of the adjacent layer where the genotype then becomes expressed, so creating a new sector. Yet another consequence of the truly mericlinal structure of the shoots is that their lateral buds frequently give rise to periclinal chimeras.

Development of periclinal chimeras

Lateral buds develop from meristematic activity in a small group of cells within the flanking meristem, and whenever this zone coincides with the periclinal part of a mericlinal chimera the developing bud is periclinal too (see Fig. 3.1;4). Plants which form few or no side branches are less likely to establish periclinal shoots than plants which branch freely. An exception is when the chimera originates so early in embryo development that the apical meristem is periclinal from the beginning. This is particularly likely to happen in plants which develop from variegated seedlings; their origin needs to be explained.

Mutations are caused by changes in the genetic material – deoxyribose nucleic acid. Most DNA is located in the nuclear genome, but plants have a smaller amount of DNA belonging to the plastid genome. The plastids are an essential group of organelles which to some extent are interconvertible and which are closely associated with a number of important functions, such as the chloroplasts with photosynthesis, the chromoplasts with the production of some red and yellow pigments, and amyloplasts with starch storage (Kirk and Tilney-Bassett, 1978). An important property of plastid genes is that they are present in many copies, which are distributed to the daughter cells at cell division in a somewhat random manner, whereas the controlled division and movement of the chromosomes ensures that each daughter cell receive two copies of all nuclear genes. A nuclear gene mutation is immediately expressed by a simple lineage in which all descendant cells are identical, whereas after a plastid gene mutation, as there are many copies of the gene within each plastid and often many plastids in each cell, there is a long delay until many divisions later when the original mutant plastid gene and associated organelle has sufficiently

Fig. 3.5 Three main shoots from a variegated seedling of *Dianthus barbatus* showing the later stages in sorting-out. At the back, the shoot has mixed and sectorially variegated leaves. On the left, the shoot is developing into a stable periclinal chimera with leaves having a narrow green margin and broad white centre. On the right, the shoot has developed into a stable periclinal chimera with leaves having a narrow white margin and broad green centre.

multiplied. During this latent period, the normal and mutant plastids of the mixed cells sort-out from one another by their chance distribution into the daughter cells at cell division. When a daughter cell has only mutant, white plastids, a pure white cell lineage is created, and likewise, with only normal plastids, a pure green cell lineage. When a daughter cell is again a mixed cell, it separates further into green or white or both types of lineage during its later divisions. Thus, as a result of a single plastid gene mutation, there are within the primary cell lineage secondary, tertiary, and many more lineages until the sorting-out of pure cells from mixed cells is complete. Hence the overall effect of the mutation is to produce a characteristic complex cell lineage chequered pattern of green and white areas (Fig. 3.5). During this sorting-out the leaf palisade and spongy mesophyll tissues appear as various shades of green, depending upon the distribution and concentration of the green, white and mixed cells. In dicots the leaves appear finely mosaiced and in monocots finely striped. The early phase of sorting-out, characterized by the finely chequered mosaic or striping, proceeds gradually, as the proportion of mixed cells decreases, to a coarse mosaic or striping. Finally, the last vestige of the sorting-out is replaced by pure green and white lineages, from which proceeds the growth of green, white, sectorial, mericlinal and periclinal shoots depending on the distribution of the green and white tissues.

In the majority of higher plants the plastids are maternally inherited (Sears, 1980). So, if sorting-out is incomplete when the egg cells are formed, the two types of plastid are transmitted into the young embryos within the seedlings of the next generation. In such cases sorting-out begins anew, in the zygote, well before the differentiation of the shoot apical meristem, which increases the likelihood of an early completion and the possibility of a periclinal chimera developing in the primary shoot. In a minority of plants the plastids are biparentally inherited. In these cases as, for example, in zonal pelargoniums, hybridization between varieties having normal and varieties having mutant plastids in their germ cells often produces some mixed zygotes that develop into variegated seedlings. Sorting-out is fairly rapid during embryo development and, when the ratio of normal to mutant plastids is approximately equal, periclinal chimeras are often formed within the first few leaves of the young primary shoot (Tilney-Bassett, 1963b). With the establishment of a periclinal shoot, vegetative propagation easily leads to a fully periclinal plant, which is usually sufficiently stable for its growth in perpetuity. Many periclinals have already been propagated for a century or two and, as there are so many, it is appropriate to now consider how they are classified.

4 Classification of Periclinal Chimeras

Major groupings

Chimeras form a large and heterogeneous group of plants, and it is convenient to classify these in three different ways (Tilney-Bassett, 1963a; Rieger *et al.*, 1968).

(i)   According to their origin   There are often authentic accounts of the origin of chimeras by spontaneous or induced mutations, by the sorting-out from variegated seedlings after plastid mutation, by grafting, by the layering of mixed populations of cells within callus tissue cultures, and by somatic hybridization through protoplast fusion.

(ii)   According to their structure   The distinction between sectorial, mericlinal and periclinal chimeras is widely used. Later in this chapter the various types of periclinal chimeras will be discussed.

(iii)   According to their behaviour   A distinction can be made between four kinds of chimera owing to the magnitude of the differences between them and what might be expected from their behaviour:
   (*a*)   *Species or graft chimeras.* These are produced by grafting together different species or genera. Just as the constituent species are likely to differ in a wide range of features, so the behaviour of their periclinal chimeras is likely to be highly variable.
   (*b*)   *Chromosomal chimeras.* These are chimeras in which the layers differ in their chromosome constitution. Occasionally chimeras arise from loss or gain of individual chromosomes or chromosome fragments owing to misdivision. More commonly cytochimeras have simple multiples of the normal chromosome complement in the changed layers. There are various effects on cell size and growth characteristics.
   (*c*)   *Nuclear gene-differential chimeras.* These chimeras arise by spontaneous or induced mutation of a nuclear gene to a dominant or recessive allele. As a rule one character is affected at a time in the leaf, flower, fruit, or other parts.
   (*d*)   *Plastid gene-differential chimeras.* These chimeras arise by spontaneous or induced mutation of a plastid gene, followed by the sorting-out of two kinds of plastid during vegetative growth. Alternatively, after selfing or hybridization, plastids may sort-out from a mixed egg or mixed zygote respectively. This type of chimera is recognized at the time of origin by the sorting-out pattern in the leaves. After sorting-out is complete,

periclinal chimeras are distinguished from similar looking nuclear gene-differential chimeras by their non-mendelian inheritance. The majority of variegated-leaf chimeras are of this kind.

All plastid gene- and some nuclear gene-differential chimeras affect the colour of the plastids within the leaves, and these are grouped together as 'chlorophyll' chimeras, or preferably as variegated-leaf chimeras. Sometimes we find chimeras with layers differing in respect of both their nuclear and their plastid genes.

In addition to these ways of classifying chimeras, they are frequently referred to by the organ affected, be it leaf, flower, fruit, or tuber. Taking the groups together, if we wanted to be pedantic, many chimeras could be described as spontaneously arising, plastid gene-differential, periclinal, variegated-leaf chimeras. In practice, it is more useful to use the different ways of classifying chimeras only as and when it is helpful to clarify the particular subject under discussion, and so reduce the risk of fitting little known plants into the wrong category.

Apical structure

Of the various ways of explaining the organization of shoot apices, the division of the apical zone into one or more tunicas and a corpus has proved the most useful for classifying periclinal chimeras. For descriptive purposes this system was simplified by Satina and Blakeslee (1941) who abbreviated the outer tunica to L I, the inner tunica to L II, and the corpus to L III. When a plant has one tunica the corpus becomes L II, and when a plant has three tunicas the corpus becomes L IV. The different genotypes within each layer are often represented, for illustrative purposes, by capital letters such as A, B, C or X, Y, Z. So, by combining layers L I to L III and genotypes A to C, the varying structure of two- and three-layered periclinal chimeras and the expected bud variations derived from them can be described and displayed in a simple manner (Kirk and Tilney-Bassett, 1978).

(i) **Two-layered chimeras**   A two-layered chimera has two genotypes for which there are two structural arrangements, either genotype A is in the outer layer L I and B in L II, or vice versa, as follows and as indicated in Fig. 4.1.

	L I	II	See Fig. 4.1 sectors
	A	B	4
or	B	A	3

Bud variations in a two-layered chimera are of two types:

(a) L I duplicates, replacing L II, so a shoot is formed composed entirely from the genotype of the skin.

L	I	II		L	I	II	See Fig. 4.2 sector
	B	A	alters to		B	B	1

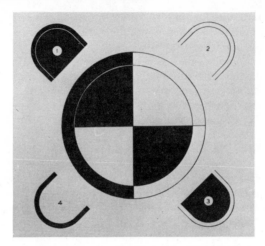

Fig. 4.1 Diagrams of transverse sections (inner circle) and longitudinal sections (outer ring) of two-layered chimeras having two structural arrangements. **(1)** Solid for genotype *AA*. **(2)** Solid for genotype *BB*. **(3)** Periclinal *BA*. **(4)** Periclinal *AB*. Black = *A*, white = *B*. (From Kirk and Tilney-Bassett, 1978.)

(b) L II duplicates, displacing L I, so a shoot is formed composed entirely from the genotype of the core.

L	I	II	alters to	L	I	II	See Fig. 4.2 sector
	B	A			A	A	2

(ii)   Three-layered chimeras   A three-layered chimera, whether it has two or three genotypes, has six possible structural arrangements.

Genetic combinations and structural arrangements of
two genotypes

L	I	II	III	See Fig. 4.3 sectors
	A	A	B	8
	A	B	A	2
	B	A	A	5
	B	B	A	6
	B	A	B	4
	A	B	B	7

Genetic combinations and structural arrangements of
three genotypes

L	I	II	III	See Fig. 4.4 sectors
	A	B	C	1
	A	C	B	4
	B	A	C	2
	B	C	A	5
	C	A	B	3
	C	B	A	6

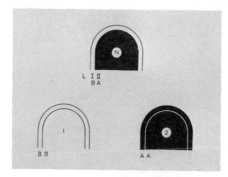

Fig. 4.2 Diagrams of longitudinal sections of apical meristems after bud variations in two-layered chimeras. **(N)** Periclinal chimera *BA*. **(1)** Solid for genotype *BB*. **(2)** Solid for genotype *AA*. Black = *A*, white = *B*. (From Kirk and Tilney-Bassett, 1978.)

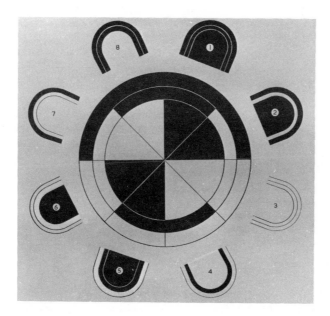

Fig. 4.3 Diagrams of transverse sections (inner circle) and longitudinal sections (outer ring) of three-layered chimeras having six structural arrangements with two genotypes giving six genetic combinations. **(1)** Solid for genotype *AAA*. **(2)** Mesochimera *ABA*. **(3)** Solid for genotype *BBB*. **(4)** Mesochimera *BAB*. **(5)** Periclinal chimera *BAA*. **(6)** Periclinal chimera *BBA*. **(7)** Periclinal chimera *ABB*. **(8)** Periclinal chimera *AAB*. Black = *A*, white = *B*. (From Kirk and Tilney-Bassett, 1978.)

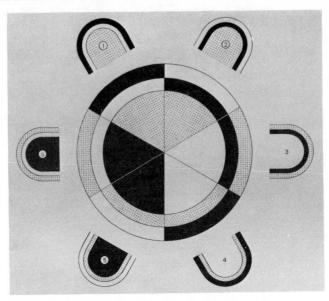

Fig. 4.4 Diagrams of transverse sections (inner circle) and longitudinal sections (outer ring) of three-layered chimeras having six structural arrangements with three genotypes. **(1)** *ABC*. **(2)** *BAC*. **(3)** *CAB*. **(4)** *ACB*. **(5)** *BCA*. **(6)** *CBA*. *Black = A*, white = *B*, dotted = *C*. (From Kirk and Tilney-Bassett, 1978.)

In the three-layered chimera with two genotypes there are six different genetic combinations (Imai, 1934, 1935c; Bergann and Bergann, 1959; Clowes, 1961) corresponding to the six structural arrangements. From the three-layered chimera with three genotypes 27 genetic combinations are obtainable; they include the three pure types and four groups of six structural arrangements (Imai, 1935c, 1937; Bergann and Bergann, 1959). These four groups comprise one with three genotypes, and three with the three combinations of two genotypes (Table 4.1). Bud variations in each of

Table 4.1 The twenty four genetic combinations attainable from a three-layered periclinal chimera with three genotypes, ABC. In addition, there are three pure types AAA, BBB and CCC.

One group Three genotypes			Three groups Two genotypes								
(A	B	C)	(A		B)	(A		C)	(B		C)
LI	II	III	LI	II	III	LI	II	III	LI	II	III
A	B	C	A	A	B	A	A	C	B	B	C
A	C	B	A	B	A	A	C	A	B	C	B
B	A	C	B	A	A	C	A	A	C	B	B
B	C	A	B	B	A	C	C	A	C	C	B
C	A	B	B	A	B	C	A	C	C	B	C
C	B	A	A	B	B	A	C	C	B	C	C

the six different structural arrangements, whether there be two or three genotypes, have nine different fates (Fig. 4.5). In each variation, periclinal division in the cells of one layer, causing duplication of that layer, leads to displacement or replacement of the cells in the adjoining layer. Displacement is a term used to describe the event in which an outer layer is lost, or shifted further outward, by duplication of an inner layer. Replacement describes the reverse event in which an inner layer is lost, or shifted further inward, by duplication of an outer layer (Dermen, 1960). The terms perforation and reduplication (Bergann and Bergann, 1959) are equivalent to displacement and replacement respectively. As a result of these bud variations, the chimeral structure alters to produce a growing-point with a new constitution.

In addition to the bud variations described, Bergann and Bergann (1962) have shown that occasionally the periclinal structure may break down to be followed by the development of a new layer arrangement, indicating that one layer has moved in relation to another layer, or two layers have changed places. Such changes appear to arise in two stages. Firstly, one of the three layers duplicates by periclinal divisions in a sector making the growing-point mericlinal. Secondly, one or two layers of the residual periclinal sector

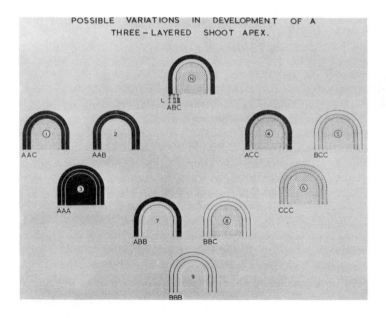

Fig. 4.5 Diagrams of longitudinal sections of apical meristems after bud variations in three-layered chimeras. **(N)** Periclinal chimera with three genotypes *ABC*. **(1)** Periclinal chimera *AAC*. **(2)** Periclinal chimera *AAB*. **(3)** Solid for genotype *AAA*. **(4)** Periclinal chimera *ACC*. **(5)** Periclinal chimera *BCC*. **(6)** Solid for genotype *CCC*. **(7)** Periclinal chimera *ABB*. **(8)** Periclinal chimera *BBC*. **(9)** Solid for genotype *BBB*. Black = *A*, white = *B*, dotted = *C*. (From Kirk and Tilney-Bassett, 1978.)

spreads sideways by anticlinal divisions pushing back the cells of the same layer in the new sector, or vice versa. In this way, by its sideways movement, one layer slides above or below another layer, or between two layers.

Imai (1934, 1935a, 1937) developed a comprehensive terminology for the types of periclinal chimera. He distinguished growing-points which were non-chimeral or homogeneous from those which were chimeral or heterogeneous. Today, the pure, non-chimeral growing-points are commonly referred to as solid, or even as homohistonts. The heterogeneous growing-points, or heterohistonts, were divided according to the position of the mutant layer. So a two-layered, or dihistogenic, chimera was either ectohistogenic, AB, or endohistogenic, BA, where A was the mutant and B the normal layer. The three-layered, or trihistogenic, chimeras were either monoheterogeneous – with one mutant layer, or diheterogeneous – with two mutant layers. Each of the six types of periclinal chimera were named by the positions of the mutant layers as ecto-, meso-, or endo-histogenic, or the three combinations of two of these together (Table 4.2). Two other terminologies, which were developed in Germany to identify three-layered chimeras with two genotypes (Hagemann, 1964; van Harten, 1978), distinguish between two monekto-chimeras, or haplochlamydeous chimeras, in which L I is distinct from L II plus L III (ABB or BAA), two diekto-chimeras, or diplochlamydeous chimeras, in which L I plus L II is distinct from L III (BBA or AAB), and two mesochimeras, in which L II is distinct from L I plus L III (BAB or ABA). In effect, these terms distinguish three types of chimera according to whether they have a thin skin, a thick skin, or a sandwich structure (Table 4.2). There are two chimeras of each type as either of the two genotypes can fill one or two layers. Imai's classification had the merit that it provided a separate name for each of the six chimeral structures, but it did not link together in pairs chimeras of a similar type; moreover, it used the histogen terminology which had already been repudiated, so it is not used now. The other two terminologies are both used.

For the numerous examples of variegated-leaf chimeras, Correns (1937) developed a specific vocabulary in which he distinguished between nuclear gene-differential chimeras – *status leucograptus*, and plastid gene-

Table 4.2 Comparison between various terminologies used in classifying three layered periclinal chimeras.

Based on layer structure

Haplochlamydeous chimera	Diplochlamydeous chimera	Mesochimera
Monekto-chimera	Diekto-chimera	Mesochimera
Thin-skin chimera	Thick-skin chimera	Sandwich chimera
ABB and BAA chimeras	BBA and AAB chimeras	BAB and ABA chimeras

Based on position of mutant layer: A

ABB: Ecto-histogenic chimera	BBA: Endo-histogenic chimera	BAB: Meso-histogenic chimera
BAA: Meso-endo-histogenic chimera	AAB: Ecto-meso-histogenic chimera	ABA: Ecto-endo-histogenic chimera

Table 4.3 Classification of variegated-leaf chimeras according to whether the mutant layer arose by nuclear or plastid gene mutation. (Adapted from Correns, 1937).

Periclinal type	Nuclear gene-differential *Status leucograptus*	Plastid gene-differential *Status albomaculatus*	Variegated-leaf structure
Thin skin	leucostephanus	albocinctus	WGG and WG
Thin skin	leucocardius	albocordatus	GWW and GW
Thick skin	leucodermis	albotunicatus	WWG
Thick skin	leucopyrenus	albonucleatus	GGW
Sandwich	meso-leucodermis	meso-albotunicatus	GWG
Sandwich	meso-leucopyrenus	meso-albonucleatus	WGW

differential chimeras – *status albomaculatus* (Table 4.3). The mesochimeras were not actually distinguished, presumably because L I usually contributes only to the epidermis so they mostly appear like the thick skinned chimeras. Correns also suggested that for the chlorotic tissue that was not white the nuclear terms could be modified to *chlorotidermis, chlorotipyrenus, chrysodermis, aureodermis* and the plastid terms to *evanidotunicatus aureocinctus* and so forth, allowing unlimited flexibility, as desired. When, as a result of genetic analysis, it was known that the plastids were inherited by both parents, Correns classified them as *status paralbomaculatus* to distinguish the plants from those species in which the plastids were inherited solely by the maternal parent – *status albomaculatus* (Hagemann, 1964). Thus a white-margined tobacco would be called *albotunicatus* and a white-margined pelargonium of the same structure, WWG, would be called *paralbotunicatus*.

Some of these terms have been used in the scientific literature with reference to particular plants, but in practice they are frequently difficult to apply. Often the origin of the chimera was not described sufficiently accurately to permit distinction between nuclear and plastid gene mutation, and the mutant tissue cannot be tested because of problems of sterility, or to its location outside the germ layer. When the origin of the chimera is known to be through a plastid gene mutation as indicated by the pattern of sorting-out and the observation of a mixed cell stage, the problem of maternal or biparental inheritance remains. Hence the *status albomaculatus* tends to be applied in a general way to all plastid gene-differential chimeras, which include many that have not been tested, as well as in a specific way to those that have been shown to exhibit a maternal inheritance of their plastids. In my view the use of some of these terms creates more confusion than clarity, especially as they cannot be accurately applied to many chimeras. Sometimes these names may prove useful, but for most purposes the use of two or three letters representing the genotypes corresponding to the two or three layers is sufficient.

Popular names

In general, chimeras do not have special names to distinguish them from other plant varieties; their cultivar names are usually well established before their chimeral nature is recognized. There is no pressing need to alter this state of affairs. In the specific case of variegated-leaf chimeras, however, the situation is rather different. They are so numerous, and usually it is so obvious that they are chimeras, that it would often be helpful if their cultivar name indicated this feature. Unfortunately, this is generally not so. There are numerous names differing in some respect from the ones used for the normal dark green leaf pigmentation of the ancestral species. Often names like *variegata, marginata, maculata, reticulata,* and *striata* give some indication of the type of foliage to expect, and prefixes like *albo-. argenteo-,* and *aureo-* give a good indication of the colour, but many names are less clear, and even the more descriptive names often do not distinguish between variegated-leaved plants that are chimeras and those that are not. Thus the term *albomarginatus (-a, -um)* frequently refer to white-margined chimeras, but margined leaves are not necessarily chimeras, and many plants that have leaves with white margins do not have *marginata* as the varietal name. In spite of this confusion, *marginata* is a useful term to indicate a white skinned periclinal chimera without specifying precisely the structure, whether two-layered (WG), three-layered and thick skinned (WWG), or a sandwich (GWG). For the reverse chimera, no term has come into common use, although Imai (1934, 1935b) suggested the term *medioalbinata* for which the likely structures would be two-layered (GW), three-layered and thick skinned (GGW), or a sandwich (WGW). Unlike the many terms in the previous section, these terms do not attempt to cover all structures, but do distinguish the most prominent and opposite types of chimera reflecting the most common form of leaf development. Non-variegated leaf chimeras seldom, if ever, have popular names indicative of their chimeral nature.

5 Analysis of Periclinal Chimeras

Morphology and anatomy

After treating seeds or shoots with physical or chemical mutagens, the induced mutations, like spontaneous mutations, are usually restricted to a single layer of the apical growing-point, so chimeras arise. There is often a delay in recognizing a chimera, and careful observation and further experimental treatment is usually needed to confirm the chimeral structure and to define its form and, when required, to recover the solid mutant. We shall therefore look briefly at methods of determining chimeral structures and of isolating the solid mutants from them; some of the examples are described and referenced more fully in the relevant chapters to follow.

The appearance of variegated-leaf chimeras usually makes their nature obvious, but in non-variegated plants this is not so. Clues to the chimera nature are frequently dependent on such features as the crumpling or rolling up of the edges of leaves, the smallness of flowers, or the sterility of anthers, which results from distortions arising from the differential growth of cells of unlike genotypes located in different layers. Whenever the growth of the core is restricted by the growth of the skin, there is a tendency for the core to occasionally break through producing distinctive patches or sectors of skin of core origin. This happens in the 'Bizzarria' orange when a citron skin appears as streaks forced out through the matrix of sour orange (see Fig. 1.5), when a greenish-yellow sector appears on the normally dark red skin of the 'Kaiser Wilhelm' apple, and when the white splashed pink spots of 'King Edward VII' tissue appear in the skin of the potato tuber 'Red King Edward'.

Histological sections may provide no clues as the genes responsible for the contrasting genotypes are often expressed in only parts of particular mature tissues. Thus in most examples of chimeras affecting the colour or texture of the skin of an apple, potato tuber, or flower, the genotype expressed is the one located in the cells of the tunica layer that gives rise to the skin. The contrasting genotype within the core tissue is not expressed, and therefore not recognizable, except where the core breaks through the surface to form the skin. Anatomical investigations are therefore of little value unless there is a clear distinction between tissues originating in different layers as with variegated-leaf chimeras. A special case is when polyploidy aids the recognition of cells derived from different layers, usually by virtue of the larger cell size and nuclear volume of the cells with more chromosomes.

Fig. 5.1 **Upper:**   The white leaf from *Pelargonium peltatum* 'L'Eléganté' has a genetically green epidermis, so the leaf has the periclinal structure *GWW*. The two green flecks on the lower margins are areas of mesophyll tissue derived from the green L I as a result of periclinal divisions occurring late in leaf development. **Lower:**   Two variegated leaves from *P.* × *hortorum* 'Flower of Spring'. **Left**, the *GWG* sandwich chimera showing several small green lobes on the edge of the white margin derived from the genetically green L I. **Right**, the apparently greed bud variation is actually a *WGG* thin-skinned chimera as revealed by the small white sector developing on the otherwise green leaf margin.

One tissue, which has proved particularly useful to examine, is the epidermis, as the patterns of cells, stomatal pores and hairs are highly characteristic. The species forming L I is usually recognized by comparing its epidermis with that of each of the two species in question, which has been very successful in the analysis of graft hybrids (see Fig. 1.1). Among variegated-leaf chimeras, examination of epidermal strips torn off the underside of leaves has proved especially useful. When the epidermis is genetically normal the guard cells lining the stomatal pores – most often found in the lower epidermis – have a few large green chloroplasts easily recognizable under the microscope. The other epidermal cells have several green chloroplasts too, but these are normally smaller than those in the guard cells and are easily overlooked at low magnifications. By contrast, when the epidermis contains genetically mutant plastids, these are colourless or yellowish, smaller than normal, and are often quite difficult to find outside the guard cells. Hence there is normally an excellent correlation between the colour and sizes of the epidermal plastids and their genotypes. Difficulty may sometimes arise when a genetically green epidermis overlies a mutant subepidermal tissue, especially if the leaves are rather old when examined, but in young leaves the colour of the chloroplasts is usually unmistakeable. Supportive evidence for a genetically green epidermis frequently occurs in the form of green flecks on the margin of white subepidermal tissue; these arise from sporadic periclinal divisions of the epidermal cells giving rise to pockets of green mesophyll tissue (Fig. 5.1). Conversely, white flecks on the rims of green leaves are indicative of a genetically white epidermis.

The existence of periclinal chimeras is often recognized by dissimilar shoots appearing on the same plant. These are larger manifestations of the small flecks, patches or sectors found on leaves, bracts, stipules, flowers or other parts. They are sometimes quite striking as, for example, pure citrus and pure orange branches on the 'Bizzarria' orange, thorny shoots on the thornless blackberries, normal fruit on giant apple sports, or the purple and yellow flowers of broom and laburnum on + *Laburnocytisus adamii*. With

Fig. 5.2   Five leaves from *Pelargonium peltatum* 'L'Elégante' showing the transition from a white-over-green chimera on the left to a completely green-over-white chimera on the right. The bud variation, often referred to as reversal, is caused by the replacement of L II and L III owing to periclinal divisions in L I at the apical meristem – *GWG* to *GGW*. (From Tilney-Bassett (1963a). With the permission of the editors of *Heredity*.)

two-layered periclinal chimeras, bud variations give rise only to the solid form of either constituent. With three-layered chimeras the possibilities are greater. A thin-skinned chimera can give rise to a thick-skinned form or to either solid form, and likewise, a thick-skinned chimera can give rise to a thin-skinned form or to either solid form. A sandwich chimera can give rise to any other form, by cell replacement or displacement, or a combination of the two, except the reciprocal sandwich structure, as well as to any solid form. Hence, among variegated-leaf chimeras, in which there is gene expression in all three layers, the sandwich structures can be distinguished from the thin- and thick-skinned chimeras by the greater variety of bud variations they produce. A very characteristic change for many white-margined, green-centred chimeras is the reversal to green-margined, white-centred chimeras (Fig. 5.2).

Adventitious buds

When propagated by root cuttings, Bateson (1916, 1921, 1926) found several plants including varieties of fancy and zonal pelargoniums that failed to come true to type. He realized that his observations were a proof of the periclinal chimeral structure of these plants, and he concluded that the adventitious buds formed on true roots arose by endogenous growth from the central tissues, which pushed through the outer cortex, and grew into plants exhibiting the characteristics proper to the core. Two of the fancy pelargoniums had been highly stable and their chimera structure was quite unexpected. Butterfield (1926) found the method useful for demonstrating the periclinal nature of the 'Cory' thornless blackberry, but not 'Austin', which was entirely thornless, and Zimmerman and Hitchcock (1951) used it to demonstrate the chimeral structure of the 'Better Times', 'Briarcliff' and 'Souvenir' roses (Fig. 5.3).

Dermen (1948) obtained adventitious buds from the outer phloem region of the 'Sweet and Sour' apple, which gave rise to the form characteristic of L III, and from a cytochimeral 'McIntosh' apple he obtained an entirely tetraploid shoot (Dermen, 1951a). Asseyeva (1927) developed the technique of excising the eyes from potato tubers to encourage the growth of adventitious buds, and thereby revealed the nature of L III. Later workers have often used the same method to help prove the chimeral structure of many potatoes. Howard (1970b) suggested that criticism of the method, owing to examples in which adventitious buds traced back to L II and not to L III, were due to faulty experimentation in which the eyes were not all removed, or were only partially removed.

Bergann and Bergann (1959) obtained adventitious buds on leaf cuttings of the zonal pelargoniums 'Freak of Nature', 'Happy Thought', 'Mrs Pollock', 'Mrs Parker', 'Wilhelm Langguth' and 'Cloth of Gold', and in every case these grew into shoots with the characteristics of the inner component, L III. By contrast, whereas Bergann and Bergann (1982) obtained a majority of adventitious shoots from L III of isolated leaves and leaf parts of the thick-skinned, GGW, and sandwich, GWG, variegated-leaf

Fig. 5.3   A specimen of the variegated-leaf chimera *Pelargonium × hortorum* 'Mrs G. Clark', which illustrates the relationship between the formation of adventitious buds on roots and the genotype of the core. In the upper figure the white centre to the green-over-white leaves is hardly visible, and the main clue to the existence of a white core comes from the white shoot growing up from below ground. In the lower figure the unravelling of the root system shows that the white shoot has developed from an adventitious bud, which reveals the genotype of the core and proves that the zonal pelargonium is a periclinal chimera with the structure *GGW*. (Adapted from Tilney-Bassett, 1963a.)

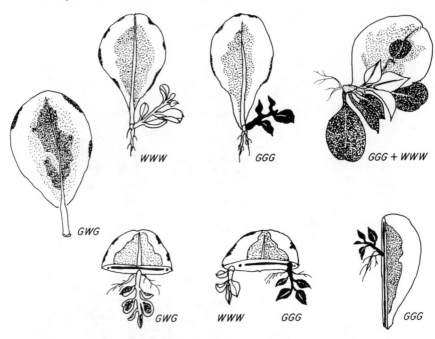

Fig. 5.4   The growth of adventitious shoots on isolated leaves and leaf parts of *Peperomia obtusifolia*. **Left**   The leaf is a typical *GWG* sandwich chimera, in which there is a white skin – with green L I and white L II – over a green core. **Upper row**   Three leaves in which adventitious buds have developed into solid white (*GWG* to *WWW*) or solid green shoots (*GWG* to *GGG*), and sometimes both types from different parts of the same leaf. **Lower row**   Transversely or longitudinally cut half leaves in which adventitious buds have developed solid white shoots from L II (*GWG* to *WWW*), or solid green shoots from L III (*GWG* to *GGG*), or occasionally have reproduced the original chimeral structure. (Adapted from Bergann and Bergann, 1982.)

chimeras of *Peperomia glabella*, *P. obtusifolia* and *P. scandens*, there was a significant minority derived from L II, plus a few variegated shoots, which must have included components from two layers (Fig. 5.4). Burk (1975) obtained adventitious shoots from leaves and leaf parts of cytochimeral and variegated-leaf tobaccos. He too found that the shoots could be derived from either L II or L III, but in this case there was no evidence for any adventitious buds containing components of both layers. It appeared that all adventitious shoots were solid having originated in a single cell, or small group of cells within the parenchyma.

In one survey over 350 plants believed to be able to develop adventitious buds on different parts of detached leaves were listed (Broertjes *et al.*, 1968), and these all subsequently developed into plantlets. The ease with which buds formed was very variable, and it was often difficult to prove that the shoots were really of adventitious origin and not from overlooked buds in or near the axils of leaves. Van Harten (1978) found that the regeneration of shoots from potato leaves, stems and stem parts was possible, but the incidence was low. The physiological requirements for adventitious bud formation depend very much on the species and variety, together with a complex of interactions between growth and environmental factors. In the genera *Achimenes, Kalanchoë, Nicotiana, Saintpaulia* and *Streptocarpus* it was shown by mutation induction, and by cytological and histological examination, that the apices of adventitious buds formed near the cut end of leaf petioles invariably originated from a single cell. The induced mutants were therefore solid, and the problem of chimeral shoots avoided (Broertjes and van Harten, 1978). Moreover, in these plants the adventitious buds were derived from an epidermal cell from L I, not from L III. In many other genera too, detached leaves and bulb scales have been successfully used for the induction of solid mutants. In other plants, such as chrysanthemum and poinsettia, adventitious buds clearly originate from more than one cell, and then mutation induction is frequently accompanied by chimerism. Various methods, tailored to the plant under consideration, have then to be devised to reduce the frequency of chimeras.

Tissue culture

On several occasions chimeras have been investigated by obtaining adventitious shoots in tissue culture. Using stem internodes of variegated periclinal chimeras of tobacco, Dulieu (1967a, 1967b) showed that the *in vitro* regenerated plantlets were derived either from L I or L II, or rarely from L III, and some new types of chimera arose with a constitution different from the original chimera, indicating a multicellular origin for at least some adventitious shoots. Opatrny and Landa (1974) induced organogenesis in cultures of isolated cotyledons from *N. tabacum* 'Burley 49', which was heterozygous for the white seedling mutant, Ws. About 1 to 2% of the regenerated plants were chimeral. When these were subcultured, they were able to regenerate both green and white shoots from different regions, and they also had some chimeral regenerants, which suggests that at least some shoots had a multicellular origin, agreeing with Dulieu's experience. Sree Ramulu *et al.* (1976a) isolated mericlinal and periclinal cytochimeras from stem explants and anther cultures of *Lycopersicon peruvianum*. In keeping with the methods of describing chimeras, it is convenient to refer to these chimeras by multiples of the basic chromosome number, x. For example, a cytochimera with the structure L I diploid, L II octoploid, L III tetraploid is abbreviated to $2x.8x.4x$, and even more simply to 284. Two cytochimeras, one with a diploid skin and tetraploid core, 244, and the second with a tetraploid skin and diploid core, 422, were then used to investigate the origin of shoots regenerated from internodes excised from stem segments (Sree Ramulu, *et al.*, 1976b). Cytological analysis showed that they regenerated from the core, L II and/or L III, and not from the skin component.

An interesting study by Cassells and Minas (1983) compared the propagation of ivy-leaf and zonal pelargonium cultivars infected with virus with cultivars having a chimeral structure. The presence of Pelargonium net vein agent, or Pelargonium petal streak agent, was proven in *P. peltatum*

Fig. 5.5 Cultures initiated from terminal buds of *Pelargonium* × *hortorum* 'J.C. Mapping', which were surface sterilized and innoculated onto medium containing Murashige and Skoog salts and vitamins, 30g l⁻¹ sucrose, 0.1mg l⁻¹ benzylaminopurine, 0.005mg l⁻¹ naphthalene acetic acid, and adjusted to pH 5.6. In addition to the white-margined shoots, typical of the parental cultivar, adventitious buds arising from the callus are developing into both green and white shoots. (Photograph kindly supplied by Dr A. Roberts.)

'Crocodile' and 'Mexicana' by graft transmission, and by the loss of their symptoms after micropropagation by bud-tip culture, whereas the viruses were maintained during explant culture. By contrast, two chimeras, *P.* × *hortorum* 'Skelly's Pride', with glossy leaves, and 'Mr Wren', with white-fringed petals edging red flowers, remained true to type after bud-tip culture, but dissociated into their chimeral components after explant culture. Chimeral breakdown also occurred after explant culture of the variegated-leaf chimeras 'Madame Salleron' and 'Mrs Cox' (Fig. 5.5).

Kasperbauer et al. (1981) found a virus infected tobacco with an unusual ruffled-leaf sector. The mutant ruffled-leaf had less leaf mid-vein and secondary vein tissue relative to the lamina than normal, which they felt might have potential value in improving the usability of tobacco leaves. A lateral shoot with a few ruffled sectors was excised, and thin layers of cells from the domes of the shoot apex and from each of nine axillary buds were excised and cultured. Of the 66 plants that were regenerated from the ten meristem cultures, two from the same explant had ruffled leaves and were virus free. The ruffled-leaf behaved as a dominant mutant. Kameya (1975) separated the components of a white-over-green pelargonium leaf in a different way. He removed the leaf epidermis, separated the remaining tissue into green and white parts, and then prepared protoplasts from each, which were suspended in culture medium. The cells developed colonies, and on transfer to a new medium these were grown into callus from which new plantlets differentiated. The plants were either green or white, and there were no variegated ones.

In vitro culture techniques are particularly valuable for the rapid propagation of plants that are normally very slow to increase by traditional vegetative means. Variegated cultivars present a special problem since now the chimeral organization needs to be conserved. Smith and Norris (1983) managed to develop a suitable medium which encouraged the development of adventitious buds on the leaf explants of three variegated-leaf chimeras of the African violet. Within six months they had obtained 80 single rosettes from ten explants of 'Calico Kitten', and an even greater yield from 'Marge Winters' and 'Bold Dance'. The authors suggested that the successful propagation of chimeras might be related to the lack of a callus intermediate stage which, when produced under the influence of growth regulators, resulted in the dissociation of the chimeral structures. Pierik and Steegmans (1983) developed a method of propagating the yellow-margined form of *Yucca elephantipes* by stimulating axillary branching, and then transferring the shoot tips into culture for further growth and rooting. Under optimal conditions, they obtained a multiplication rate of 7 to 8 shoots in three months with safe transfer to soil.

Seedlings

It is a useful rule that the germ cells, which after meiosis, fertilization, seed development and germination give rise to the seedlings, normally develop from sub-epidermal cells. In flowering shoots these sub-epidermal cells are usually derived from L II, and so the appearance of the seedlings, particularly after selfing, has become a powerful indicator for L II. This has been especially useful in the analysis of variegated-leaf chimeras. Thus, on selfing white-margined zonal pelargoniums, Baur (1909a) obtained wholly white seedlings indicating that the germ layer corresponded with the outer, white skin of the leaf. It was later shown that the white skin of many variegated zonal pelargoniums was bounded by a genetically green epidermis derived from L I; these pelargoniums were *GWG* sandwich chimeras. By contrast, when seedlings were raised from the two variegated forms of *Hosta fortunei*, they were green from the white-over-green form (Chodat, 1919) and white from the green-over-white form (Collins, 1922). Identical results were obtained from the variegated forms of *Hosta lancifolia* (Yasui, 1929). Hence these monocots were two-layered with the L II germ layer corresponding to the central core tissue of the leaf, and not the skin. A dicot that was also two-layered and with similar behaviour to *Hosta* was the white-margined *Veronica gentianoides* (Correns, 1920); in this case too the white skin appeared to be derived from L I and the green core from L II (Chittenden, 1927).

A problem that may sometimes arise was encountered by Chittenden (1926) when he obtained green and white seedlings from the green lobed, *GWG*, and yellow and white seedlings from the yellow lobed, *YWG*, cultivars of *Hydrangea macrophylla* 'Variegata'. Evidently the ovules were able to develop in tissue of both L I and L II origin. This does not mean that ovules developed in a variable position. What it does mean is that the placental initials, which are usually derived from L II, were frequently replaced as a result of periclinal divisions in L I. Consequently, when the placentas developed in their normal position, they sometimes had to develop from initials derived from L I and so produced seedlings corresponding to L I. An even more atypical result occurred with the zonal pelargonium 'Golden Brilliantissima', which was a white-over-green chimera with yellow lobes, *YWG*. In this case the few seedlings were all yellow, and the expected white seedlings absent (Neilson-Jones, 1934; Bergann, 1962b). Either L I always replaced L II during the development of the sporogenous tissues of this plant, or possibly the potentially white seedlings were selected against through being too weak to germinate. This might also be the explanation for green seedlings from *H. macrophylla* 'Nivalis', which was a thin-skinned chimera with green L I and white L II and L III, and from the green flecked white shoots of *Pelargonium* × *hortorum* 'Freak of Nature' (Chittenden, 1926; Bergann, 1962b). The lesson is to interpret the evidence from seedlings in the light of other information and not to adhere blindly to the L II rule; exceptions do occur. In peach, seed progeny helped elucidate the nature of the three kinds of variegated leaves (Foyle and Dermen, 1969), and many other variegated-leaf chimeras have been tested similarly, especially in the case of plastid gene mutants in association with tests for maternal or biparental plastid inheritance in over fifty species (Kirk and Tilney-Bassett, 1978; Sears, 1980).

The progeny of various kinds of potato chimera were analysed to help unravel the structure of leaves and tubers in many cultivars, and seed progeny have proved useful in analysing cytochimeras and many kinds of flowers and fruit chimeras. The analysis of the progeny formed part of the investigation into the possibility of chimeras for an acidless factor in orange and lime cultivars (Cameron and Soost, 1979), an early maturing factor in the Japanese satsuma mandarin (Nishiura and Iwamasa, 1969), and as a probe into the possible cytochimeral nature of the 'Centennial', 'Ishihara' and 'Red Pearl' grape vines (Yamane *et al.*, 1978).

Irradiation

Various kinds of irradiation treatment have been used to induce mutations in cultivated plants, and this had often led to the formation of chimeras (Broertjes and van Harten, 1978). X-rays, and to a lesser extent γ-rays, have also been used to alter the structure of periclinal chimeras, and to encourage the growth of solid mutants from them. Asseyeva (1931) used X-irradiation to alter a thin-skinned potato tuber with L I mutant to a thick-skinned tuber with both L I and L II mutant. Similarly, irradiation of 'Red King Edward', a bud sport of 'King Edward VII' potato, in which L I is genetically red instead of white splashed pink, transformed it into a periclinal chimera (*RWW* to *RRW*) in which both L I and L II were genetically red (Howard, 1959, 1964b). Other potato tubers chimeral for skin colour and altered by irradiation treatment are referred to later in the book, and also potato plants altered by leaf mutations. In most of these cultivars the chimera was altered to the solid form for both alternative skin colours or leaf shapes, as well as to the thick-skinned chimera form, where this could be identified. Generally speaking, a range of irradiation doses was used up to a maximum of 4 krad. The results were variable and difficult to interpret. In practice, increasing the dose increased the frequency of change and increased the frequency of solid tubers formed. Moreover, while there were differences between experiments – probably arising at least partly from variation between cultivars – the solid tubers developed more often out of the core tissues L II + L III than out of the skin tissues. For example, after a dose of 3 krad of X-rays, Klopfer obtained about 10% pure red tubers from 'Rote Holländische Erstling' (RYY to $RRY + RRR$) and 25% yellow tubers (RYY to YYY); the remainder were unaltered. Van Harten (1978) suggested that within genetically homogeneous plants the differences in radiosensitivity between layers was probably caused by differences in physiological or mitotic activity. He also emphasized the difficulty in estimating the real activity of each layer owing to sources of error from underestimating the frequency of the change from a thin-skinned to a thick-skinned chimera (YRR to YYR), which is difficult to identify as the genotype is not expressed in all tissues, and from overestimating the duplication of L II + L III (YRR to RRR) owing to regeneration from a large reservoir of subepidermal cells from within the real apical zones as well as from outside this area.

The carnation 'William Sim' had red petals; two clones derived from it had a mutant L I, which gave rise to white or pink petals. An irradiation dose of up to 5 krad X-rays destroyed the outer layer of the periclinal chimera, which was displaced by regeneration of a new epidermis from the deeper red cells (WRR to RRR). The change from white, or pink, to red in 50 to 90% of flowers was too frequent for mutation, whereas the reverse change from red to white, or pink, in 4% of flowers, most probably did arise by mutation (Sagawa and Mehlquist, 1957). By contrast, an orange red-striped chimera of carnation, with a genetically yellow L I, gave a higher frequency, after 5 krad and 7 krad of γ-rays, of solid yellow flowers (YRR to YYY) than red flowers (YRR to RRR). Péreau-Leroy (1969, 1974) explained these results by assuming that the high dose rates caused more damage to the inner than the outer layers.

Preferential multiplication of L I rather than L II or L III was found after treating with 3 krad X-rays a periclinal chimera affecting the colour of bracts in *Euphorbia pulcherrima* 'Eckes Rosa' (Pötsch, 1966a), whereas regeneration from L III was more common after irradiating leaf chimeras in *Abutilon × hybridum* (GWG to GGG) and *Pelargonium × hortorum* 'Saleron BBC' (BBC to $BCC + CCC$) (Pötsch, 1966b, 1969).

Gamma-irradiation of growing trees or dormant scions of colour sports of 'Delicious' and 'Rome' apples induced reversions in fruit colour from blushed to striped or to a less red colour (Pratt *et al.*, 1972b). As most of the anthocyanin-containing cells producing the original red colour of the sports occurred in the hypodermis, derived from L II, it was assumed that these were periclinal chimeras for a mutation increasing anthocyanin content in L II or, by layer replacement, in both L II and L III. Irradiation treatment was also used to investigate apple russet sports (Pratt, 1972).

Experiments with cytochimeral apples showed that X-irradiation of shoot apices was a useful means of altering the layer structure so that, for example, a $2x/4x$ periclinal chimera with a diploid L II might be altered to one with a tetraploid L II owing to the change 224 to 244 (Pratt, 1960). Pratt (1963) exposed dormant apple scions to 3 krad of X-rays or to 3 h of thermal neutrons and grafted to untreated stocks. After damage to the central apical cells, shoot growth resumed either by the growth of adjacent cells which reformed the meristem, or by cells lateral to the damaged area forming a substitute meristem, or by the development of an axillary meristem developed at one or more nodes below the apical meristem. The chromosome constitution of the recovered shoots depended on the ploidy of the cells taking part and their subsequent arrangement into layers. Generally speaking increase in $2x$ outer tissues (2244 to 2224) was more frequent than increase in $4x$ inner tissues (2244 to 2444), but the direction of change appeared to be partly related to the length of the shoot and to the apple cultivar used. After irradiation, the cytochimeras included new types that had not been found by branch selection (Pratt *et al.*, 1967), which proved the value of irradiation for increasing the variability compared with untreated lines.

6 Cytochimeras

Colchiploidy

Whenever cytologists discover plants in which the cell population contains cells varying in chromosome number, whether euploid, or aneuploid, the plants may be described as mixoploid (Nemec, 1910), a term which embraces all types of mosaics and chimeras so long as there is heterogeneity in the chromosome constitution. The finer mosaics, which largely result from aneuploidy, arising from unequal chromosome segregation after interspecific hybridization, protoplast fusion, or regeneration from callus cultures, do not usually develop into stable chimeras; they are mostly of interest to students of chromosome behaviour. Our interest lies in those mixoploids, which result largely from polyploidy, in which there is a clear development into a sectorial or mericlinal chimera, and eventually into a stable periclinal chimera. These plants, which Dermen and Bain (1944) called cytochimeras, have proved to be of great importance for our understanding of chimeras.

All genera or families of flowering plants with basic chromosome numbers of $x = 12$ or higher, as well as some with $x = 10$ or 11, have probably been derived by polyploidy (Stebbins, 1971). Hence polyploidy is widespread and very significant, and if this process is so successful in nature, may it not also prove attractive if man could raise polyploids artificially? The increase in gene dosage resulting from the multiplication of the chromosome sets invariably increases cell size and often the size of many plant parts. The improvement of many agricultural and horticultural crop plants by increasing flower size and fruit and seed weight is a particularly attractive prospect. Moreover, it may not matter that polyploids frequently have disadvantages initially, for example, a reduction in resistance to drought and cold and lower fertility and yield, as this can probably be overcome by further breeding and selection at the polyploid level. Accordingly, it was an exciting and highly significant advance when Blakeslee and Avery (1937) found that the water soluble alkaloid colchicine could be used to induce polyploidy, artificially. In nature polyploidy usually arises through the union of unreduced gametes (Harlan and de Wet, 1975), whereas colchicine inactivates the spindle of a dividing cell and arrests mitosis at metaphase (Sharma and Sharma, 1972). The chromosomes divide, but with no spindle daughter chromatids cannot be drawn to opposite poles, so they remain at the equator of the cell, where they become enclosed in a common nuclear membrane instead of separating into two daughter cells. If meristematic cells have prolonged contact with colchicine,

Table 6.1 List of higher plant species that have developed periclinal cytochimeras either spontaneously, or after various kinds of induction, and an indication of the relative frequencies of the various types.

Species	Cultivars	Origin	Thin skin		Thick skin		Sandwich		Investigators
			244*	422*	224	442	242	424	
Alnus glutinosa	—	C	+	+	+	—	—	—	Johnsson, 1950
Amaranthus caudatus	—	C	11	6	—	—	7	2	Behera *et al.*, 1974
Ananas comosus	Cayenne	C	+	+	—	—	—	—	Kerns and Collins, 1947
Bougainvillea spectabilis × *B. peruviana*, and	Thimma	S							Ohri and Khoshoo, 1982
B. glabra × *peruviana*	Mrs.McCleans	S							
Brassica oleracea	Ferry's Hollander	C							Newcomer, 1941
Camellia lutchuensis × *C. rusticana*	Fragrant pink	C	+	—	+	—	—	—	Ackerman and Dermen, 1972
Citrus deliciosa × *tangerina*	King × Dancy	S	1	—	—	—	1	—	Frost and Krug, 1942
Citrus spp., *grandis*, *paradisi*, *recticulata* and *sinensis*	20 named cultivars and hybrid clones	C	13	4	4	6	4	2	Barrett, 1974
Citrus sinensis × *Poncirus trifoliata*	Carrizo	S	+	—	—	—	—	—	Barrett and Hutchinson, 1982
Coffea canephora	—	C	+	—	—	—	—	—	Noirot, 1978
Crepidiastrixeris denticulatum	—	S							Humihiko, 1941
Datura stramonium	—	C	10	30	10	—	4	4	Satina *et al.*, 1940
Fortunella margarita	—	C	—	—	—	—	1	—	Barrett, 1974
Hemerocallis sp.	24 named cultivars	C	2	7	—	13	4	—	Arisumu 1964, 1972
Lilium longifolium	—	C	+	+	—	—	—	—	Emsweller 1947, 1949; Emsweller and Stewart, 1951
Lycopersicon peruvianum	—	T	+	+	—	+	—	—	Sree Ramulu *et al.*, 1976a, b
Malus pumila var. *domestica*	(Table 6.2)								
Morus alba	Ichinose	G	53	5	2	—	7	1	Katagiri, 1976; Katagiri and Nakajima, 1982

Species	Cultivar(s)	Origin							Reference
Pisum sativum	—	C	—	—	—	1	—	—	Straub, 1940
Prunus persica	Elberta, and Golden Jubilee	C	—	+	+	—	—	—	Dermen, 1947b, 1953b; Dermen and Stewart, 1973
Pyrus communis	Improved Fertility, and Double Williams	S	2	—	—	—	—	—	Marks, 1953
Ribes nigrum × grossularia	Boskoop Giant × Early Sulphur	C	+	—	—	—	—	—	Nilsson and Goldschmidt, 1962
Rosa multiflora	—	C	+	1	—	—	—	—	Semeniuk and Arisumi, 1968
Solanum bulbocastanum	—	C	9	3	—	—	—	—	Hermsen and de Boer, 1971
Solanum × juzcpczukii	—	C	—	—	—	—	—	+	Howard *et al.*, 1963
Solanum tuberosum	—	C	+	+	+	—	—	—	Baker, 1943; Klopfer, 1965a; Frandsen, 1967
Sorghum vulgare	—	C	—	+	—	—	—	—	Siddiq, 1967
Vaccinium oxycoccus	—	C	+	+	+	—	—	—	Dermen and Bain, 1951; Dermen, 1945
Vitis rotundifolia	Higgins	G	+	—	—	—	—	—	Fry, 1963
Vitis vinifera	Portland, Wallis Giant, and Starks Early Giant	C	+	—	—	—	—	—	Dermen, 1954
Vitis vinifera	Eaton, and Fredonia	S	2	—	—	—	—	—	Einset and Lamb, 1951a
Vitis vinifera	8 named cultivars	S	+	—	—	—	—	—	Einset and Pratt, 1954
Vitis vinifera	62 named cultivars	C/S	45	2	—	—	—	—	Thompson and Olmo, 1963
Vitis vinifera	Portugieser, Riesling, and Sheurebe	S	3	—	—	—	—	—	Staudt, 1973

Key: C = Colchicine-induced; G = Gamma irradiation-induced; S = Spontaneous origin; T = Tissue culture origin.
+ = Chimera found; − = Chimera not found. Blank: Chimeras reported but not classified.
* Where only thin-skinned chimeras are reported, the growing-point might be only two-layered. Strictly speaking, if any plants were initially tetraploid, or doubled twice, the series should be 488, 844, 884, 448, 484 and 848. Occasionally, as in *Datura*, there may also be chimeras of three ploidy levels.

the chromosomes may pass through a further cycle of DNA synthesis and replication before once again becoming arrested at the next metaphase. So in the first round of chromosome division diploid cells become tetraploid, and in the second round the tetraploid cells, in turn, become octoploid.

When polyploidy is induced in an apical meristem, it affects cells in one, two, or all three layers and so gives rise to a mericlinal shoot from which a periclinal cytochimera or polyploid cell lineage may subsequently develop. A notable feature of the colchicine-induced polyploidy in *Citrus* species was the high frequency of the occurrence of adjacent layer changes. Barrett (1947) found among the 20 clones treated, plus *Fortunella margarita* (Table 6.1), that of those affected a single layer change occurred in 19.4%, a two layer change in 28.4% and a three layer change in 49.2% of the shoots isolated. Hence the response was not consistent with the probability concept of mutation, in which the chance of change in two or three layers is equal to the product of the probability of change in individual layers.

After treating the thorn apple with colchicine, Blakeslee and collaborators obtained five of the six possible diploid-tetraploid cytochimeras, and several combinations of diploid-octoploid, tetraploid-octoploid, and even a diploid-tetraploid-octoploid cytochimera (Blakeslee *et al.*, 1939, 1940; Satina *et al.*, 1940). Following the success of Blakeslee's group, many later workers have continued the use of colchicine to induce polyploidy – a method which was later to be referred to as colchiploidy (Dermen, 1954) – and many of their efforts led to the incidental development of periclinal cytochimeras.

Induction and recognition of cytochimeras

Colchicine has usually been applied to plant tissues as a solution in water at a concentration varying between 0.01 and 1%. Sometimes glycerine and wetting agent are added. Less frequently colchicine is made up as a paste mixed with lanolin or glycerine. Noirot (1978) used 1% colchicine in an emulsion of equal parts lanolin and coconut milk. Seeds of the thorn apple (Satina *et al.*, 1940), potato (Baker, 1943; Frandsen, 1967) and aubergine (Wanjiri and Phadnis, 1973) were soaked in colchicine for periods from one day to over a week, whereas in the barley (Greis, 1940), cabbage (Newcomer, 1941), pineapple (Kerns and Collins, 1947), potato (Baker, 1943), *Ribes* hybrids (Nilsson and Goldschmidt, 1962), *Solanum bulbocastanum* (Hermsen and de Boer, 1971), and *Sorghum vulgare* (Siddiq, 1967) germinated seed or seedlings were used.

Often the method of application has depended very much on the nature of the plant. Colchicine was applied to detached bud scales of lily (Emsweller, 1947), to pineapple crowns (Kerns and Collins, 1947), to rooted cuttings of *Camellia* (Ackerman and Dermen, 1972), and to cuttings, slips and suckers of *Ribes* hybrids (Nilsson and Goldschmidt, 1962). Dermen and Bain (1944) treated the lateral buds of cranberry, as did Semenuik and Arisumi (1968) with *Rosa multiflora*, and Dermen (1954) concentrated his treatment on the upper buds of grapes. Frequently, the shoot apices were treated as in *Amaranthus* (Behera *et al.*, 1974), daylilies (Arisumi, 1964), peaches and pears (Dermen, 1947b, 1947c) and the pea (Straub, 1940). As a rule colchicine was applied by surrounding buds and apices with a cotton plug soaked in the solution, and adding further drops of solution at intervals to ensure that the tissues remained moist; usually there were at least three applications spread over one to several days. As colchicine is rather toxic overdosing kills the tissues, and so the ideal dose for the plant in question had to be found by trial and error.

Following the colchicine treatment, many workers have carefully pruned the trees and bushes in order to encourage the growth of treated buds and shoots, and to discourage or stop growth from untreated parts. Furthermore, when the treated organs grew, they were often pruned too, so as to encourage the growth of shoots and sides of shoots wherever polyploid induction appeared to have been successful, and to suppress unaltered growth. In this way any success with the initial colchicine treatment was greatly enhanced by the skill of the experimenter in recognizing and selecting the new polyploid areas and encouraging their growth over a period of one to several years.

Partly owing to the toxicity of colchicine and the consequent death of some cells, and partly owing to the induction of polyploidy in others, the regular development of treated shoots is invariably disturbed. Precisely what form the disturbance takes depends on the species. In general, shoot growth is retarded, stunted and twisted, as was the case in *Amaranthus* (Behera *et al.*, 1974), *Camellia* (Ackerman and Dermen, 1972), grapes (Dermen, 1954) and roses (Semenuik and Arisumi, 1968). The stems and particularly the leaves of *Amaranthus, Citrus* (Frost and Krug, 1942). cranberry (Dermen and Bain, 1944), daylilies (Arisumi, 1964), *Ribes* hybrids (Nilsson and Goldschmidt, 1962) and roses were described as thicker than normal, and they were often broader (Fig. 6.1) as in *Coffea canephora* (Noirot, 1978) and *Solanum bulbocastanum* (Hermsen and de Boer, 1971), and curled at the edges. Sometimes there are diagnostic changes in morphology. The polyploid leaves of muscadine grape vines developed a striking U-shape to the leaf blade (Fig. 6.1) in the region of the petiole attachment (Dermen, 1954), and the leaves of octoploid potatoes were more glabrous, succulent and brittle than the leaves of the normal tetraploid potatoes (Baker, 1943); the leaf serrations of *Camellia* were deeper and the marginal teeth more pronounced (Ackerman and Dermen, 1972). In *Amaranthus* (Behera *et al.*, 1974), *Camellia, \Citrus* species (Barrett, 1974) and grapes (Dermen, 1954), particular attention was drawn to the clear mosaic or sectoring created by the contrast between the normal green diploid tissues and the darker green tetraploid tissue. Although less important than the vegetative growth for the initial recognition of polyploid areas, the flowers of cranberry (Dermen and Bain, 1944) and grapes (Dermen, 1954) were enlarged by polyploidy, as were the fruit, but, as the fertility of the autotetraploids was much reduced, there were fewer fruit. Large sized, or 'Giant', table grapes are generally considered more attractive and fetch higher prices than smaller ones. So the development of polyploids by colchicine treatment seemed to be a promising way of

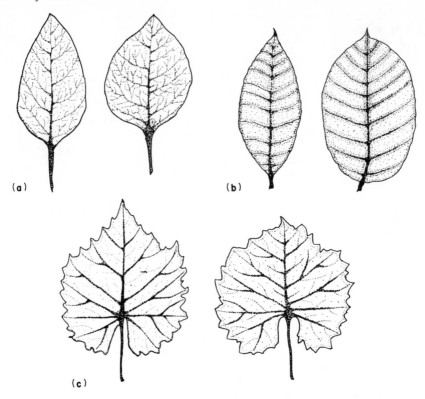

Fig. 6.1 Illustrations to show the contrast between the longer, narrower shaped leaves of diploid species on the left, and the slightly shorter, broader shaped leaves of their tetraploid derivatives on the right. **(a)** *Solanum bulbocastanum*; **(b)** *Coffea canephora*, **(c)** *Vitis rotundifolia*. (Adapted from Dermen, 1954; Hermsen and de Boer, 1971; and Noirot, 1978.)

improving suitable varieties. In the pineapple, polyploids flowered later, fruit maturity was delayed, and overall fruit size reduced even though the individual fruitlets (or eyes) are larger (Kerns and Collins 1947). Internally, the pineapples had a lower sugar content, a higher percentage of water, and a lower dry matter content in the polyploid than the corresponding diploid variety. Clearly there was a need for much further improvement through breeding and selection before these polyploids could be of commercial value.

After the induction, selection and propagation of apparently polyploid shoots and plants a more detailed analysis is needed to determine the precise nature of the end-product. In short, to find out how successful has been the selection, which assumes a strong association between polyploidy and the observed structural deformities and morphological changes just described. A doubling of the chromosome complement of a cell is bound to increase the nuclear volume, and this is invariably associated with an increase in cell

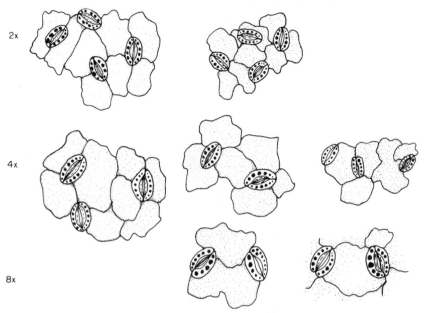

Fig. 6.2 Illustrations of stomata among epidermal cells to show the effect of increasing ploidy upon cell size. Left cranberry, centre thornapple, right potato. Upper diploid, middle tetraploid, lower octoploid. Magnification × 175. (Adapted from Dermen and Bain, 1944; Satina *et al.*, 1940; and Baker, 1943.)

size. Hence the identification of polyploid tissues is based on nuclear size, cell size and, where possible, chromosome counts. One type of tissue which is very simple to examine is the leaf epidermis. A fragment of the lower epidermis is easily torn off the leaf blade and examined under the low power of the compound miscroscope. In numerous such observations the clearly distinguishable stomatal cells of a polyploid epidermis are significantly larger than those of the corresponding diploid cells (Fig. 6.2). Moreover, the well developed chloroplasts of the stomatal guard cells increase in number with increasing ploidy levels (Butterfass, 1979). Hence the ploidy level of L I is quickly determined. Another useful cell, which is accessible at flowering, is the pollen grain. Diploid pollen grains derived from tetraploid germ tissues are significantly larger than the normal haploid pollen grains derived from their diploid germ tissues. As these pollen grains are derived from the pollen mother cells, which develop in the subepidermal layer of the anthers, they provide the key to the constitution of L II. Further evidence of the ploidy level may be obtained by observing the chromosomes – univalents, bivalents, multivalents – at pollen mother cell meiosis, or at pollen grain mitosis. One may also deduce much about the chromosome complement of the germ layer after selfing and intercrossing the plants, and examining the chromosome constitution of their progeny by standard root tip squashes. Ascertaining the nature of L III is more difficult as it involves the development of adventitious buds derived from the inner tissues, or as a result of propagation by root cuttings. Alternatively, useful deductions can sometimes be made about L III from the range of cytochimeras available and the bud variations to which they give rise. Finally, as shall be seen later, the ploidy level of all layers can be determined from differences in cell size, nuclear size, and chromosome number in thin longitudinal sections through the growing-points of apical or lateral buds.

An alternative method to the use of colchicine for inducing polyploidy is to irradiate shoot tips with γ-rays derived form a ^{60}Co source. Fry (1963) irradiated muscadine grapes with a dose of approximately 175 roentgens per hour to give a total dose of 2350 to 4350 roentgens, and Katagiri (1976) and Katagiri and Nakajima (1982) irradiated mulberry with 5 krad either over a period of one hour, or over 25 hours at a rate of 0.2 krad per hour. In both cases irradiation caused considerable cell damage, later followed by cell and tissue recovery. The particular response appeared to depend on many factors such as the irradiation dose and dose rate, the state of growth of the treated material, the position of the buds on the irradiated shoots, and the effects of pruning. During cell and tissue recovery the occurrence of polyploid cells in the recovered meristem did not appear to be random, which suggests that the cells of one layer might be more sensitive to radiation than the cells of another layer, and that recovery might be more likely from a flanking than from a central meristem but, as with colchiploidy, the frequency of the different types of cytochimera isolated is most certainly affected by the selection processes as well. After irradiation, as after colchicine treatment, the initial selection of polyploid shoots was visual followed by a more detailed cytological analysis.

Polyploid cytochimeras

A list of cytochimeras (Table 6.1) shows that most species in the table have developed at least one, and often both, of the two thin-skinned chimeras, whereas the thick-skinned and the sandwich chimeras are less well represented. One reason for the universality of the thin-skinned types is that some species develop leaves from only two layers – from a single tunica and corpus – and so are unable to form the three-layered chimeras. This is quite likely to apply to the monocots pineapple, easter lily and millet, and to some dicots like the grapevine. In coffee, the alder, rose and currant hybrid, the restriction to thin-skinned chimeras is possibly a function of the sample size; it would be premature to conclude that they have two-layered apices. As to the relative frequency of the different cytochimeras, it is probable that unconscious selection has played a part. Whereas tetraploidy in the single cell thick epidermis has little effect on leaf form and growth, tetraploidy within the inner tissues is responsible for those modifications so characteristic of polyploidy. In this respect, Frost and Krug (1942) considered that the difference between the diploid-like 242 cytochimera and the tetraploid-like 244 cytochimera was an indication of a major role for L III in determining the growth habit of *Citrus* leaves. By contrast, Arisumi (1964, 1972) concluded that the major and really significant change in the morphology of day-lilies occurred when L II was tetraploid. No doubt the difference between these two plants reflects a major role for L III in *Citrus* and for L II in *Hemerocallis* particularly in the contribution they make to the leaf tissues. In spite of the insignificant effect upon growth of a tetraploid L I, 422 cytochimeras may be picked up during the period following treatment when the shoots are closely examined for altered sectors, especially owing to frequent examination for enlarged stomata. In addition to these suggestive explanations for the variability in cytochimera formation, it may be argued that some layers are inherently more likely to become polyploid than others, although the limited data do not show any obvious trend.

Following the success in inducing polyploidy, and the finding that besides wholly tetraploids there were many cytochimeras, it was only natural that a closer look should be given to some of the existing tetraploids that were propagated vegetatively. A few of these did indeed turn out to be cytochimeras, mostly with a diploid epidermis (Table 6.1). Of course, one does not expect to find cytochimeras among tetraploids raised from seedlings. Among apple varieties, as a result of the close examination of 'large' and 'giant' sports, three main types of cytochimera were found, which Einset *et al.* (1947) classified according to the thickness of the diploid skin as type I – 2444, type II – 2244, and type III – 2224 (Table 6.2). The classification made the rare assumption that, at least for a significant period of development, the growing-point of the apple shoot had three tunica layers plus a corpus contributing to leaf and fruit development. These cytochimeral trees were vigorous growers; the largely tetraploid type I was more vigorous than the diploid parents, and the progressively less tetraploid types II and III about the same as their parents. All the sports tended to be flatter-shaped trees owing to wider angles between the branches and trunk. Other vegetative features included some typical characteristics of cytochimeras – the leaves were rougher and darker than those of their parents, the petioles were thicker and the buds were generally broader. The trees bloomed at the same time and were often less productive than their parents. Except for 'Ontario', the sports were selected as large-fruited forms. The sports of 'Jonathan', 'Northern Spy' and 'Wealthy' developed large, symmetrical apples which were more flattened and with a broader basin than normal, the calyx lobes were broader and fleshier. The 'Rome' sport had fruit which were outstanding because of their large size and irregular, knobbly form. Indeed, it was this sport with the most diploid and least amount of tetraplod tissue that appeared to differ most from the parental form (Einset *et al.*, 1947). An irregular, radially unsymmetrical giant sport of the 'Canada red' apple might have been a cytochimera, but alas the necessary tests were not made (Gardner *et al.*, 1948).

The large and giant sports had been considered as mere curiosities as the trees were often unproductive of fruit and the irregular fruit shape displeasing. Moreover, they had the reputation of being coarse, and earlier maturing than normal (Darrow *et al.*, 1948). Nevertheless, these pomologists began to realise that some of the cytochimeras had great potential in breeding for the mass production of polyploid apples, from

Table 6.2 List of apple cytochimeras derived from spontaneous polyploidy. Classified on the assumption of a four-layered growing-point into three main and one additional types; the original growers or sources of the cytochimeras are given in brackets.

Type, cultivar and source	Investigators
Type 1 – 2444	
Delicious	Dermen, 1952b, 1955
Golden Delicious (Doud)	Pratt et al., 1967
Giant Jonathan (Conkle)	Einset, 1952
Giant McIntosh (Kimball)	Dermen and Darrow, 1948; Dermen, 1951a, 1952a, 1955; Pratt et al., 1967
Giant Wealthy (Loop)	Einset et al., 1947; Darrow et al., 1948; 1952; Einset, 1948, 1952; Pratt et al., 1967
Large Wealthy (Stevenson)	Einset et al., 1947; Einset, 1948, 1952; Pratt et al., 1967
Large Yellow Transparent (Perrine)	Einset and Imhofe, 1949; Einset, 1952; Pratt et al., 1967
Ontario Sport (Geneva)	Einset et al., 1947; Darrow et al., 1948; Einset, 1948, 1952; Einset and Imhofe, 1949; Dermen, 1955; Pratt et al., 1967
Wrixparent (Pat. No. 388)	Einset and Lamb, 1951b; Einset, 1950, 1952; Dermen, 1955; Pratt et al., 1967
Type 2 – 2244	
Giant Jonathan (Adams)	Einset and Imhofe, 1951; Einset, 1952
Giant Jonathan (Edwards)	Einset and Imhofe, 1951; Einset, 1952
Giant Spy (Loop)	Einset et al., 1947; Darrow et al., 1948; Einset, 1948, 1952; Pratt et al., 1967
Grimes Sport (McClintock)	Einset and Imhofe, 1949; Einset, 1952
Jonathan Sport (Welday)	Einset et al., 1947; Darrow et al., 1948; Einset, 1948; 1952; Pratt et al., 1967
Large Wealthy (Coombs)	Einset and Imhofe, 1949; Einset, 1952; Pratt et al., 1967
Large Yellow Transparent (Perrine)	Einset and Imhofe, 1949; Einset, 1952
McIntosh Sport (Robinson)	Darrow et al., 1948; Einset and Lamb, 1951b; Einset, 1952
Winesap	Dermen, 1951a, 1953a, 1955
Type 3 – 2224	
Bates Lobo	Pratt, 1963; Pratt et al., 1967
Giant Rome (Loop)	Einset et al., 1947; Einset, 1948, 1952; Pratt et al., 1967
Golden Delicious (Perrine)	Pratt et al., 1967
Large McIntosh (Cornwall)	Einset and Imhofe, 1949; Einset, 1952
Additional type – 2422	
Giant Jonathan (Conkle)	Einset, 1952; Pratt et al., 1967
Giant McIntosh (Kimball)	Dermen, 1948, 1951a, 1955; Dermen and Darrow, 1948; Pratt et al., 1967
Levering Limbertwig	Pratt et al., 1967
McIntosh (Johnson)	Pratt et al., 1967

which selection for desirable seedlings could be made on a large scale. Einset and Lamb (1951b) assembled chromosome counts for apple varieties and found 163 diploid, 11 triploid, 10 cytochimeras and no wholly tetraploid apples. In order to appreciate the significance of these figures, Einset (1948, 1952) determined the ratio of the spontaneous occurrence of triploids from diploid parents as about 1:350, and the occurrence of tetraploid as about 1:1100, whereas tetraploid seedlings arose from triploid parents in a ratio of about 1:40. It thus appeared that triploid varieties were more frequent than one would expect from the frequency of their spontaneous occurrence as triploid seedlings. Evidently the triploids often had valuable qualities for which they had been strongly selected. By contrast, the tetraploids had made no impression partly, perhaps, because of their rarity, and partly, no doubt, because without a better source than spontaneous polyploidy there had been too little opportunity, through breeding and selection, to upgrade any initial tetraploids to viable commercial products.

As the type I cytochimeras had a tetraploid subepidermis, they were expected to behave as tetraploids in breeding, and therefore could be used as a source of new tetraploid seedlings, after selfs or intercrosses, or as a source of triploid seedlings, after crosses with diploid varieties (Darrow et al., 1948; Dermen and Darrow, 1948; Einset, 1948). The other types of cytochimera could be used if the tetraploid tissues were isolated from them, or if they were relocated into the subepidermis. Dermen (1948, 1951a, 1955) obtained tetraploid shoots from the cytochimeras 'Delicious', 'Giant McIntosh', 'Ontario Sport', 'Winesap' and 'Wrixparent' by a technique in which he encouraged the formation and growth of adventitious buds on shoots in which the natural buds had been removed. Another technique, used by Pratt (1960, 1963), was to irradiate dormant scions with a dose of 3 krad X-rays, or to treat them for 3 hours with thermal neutrons, and then to graft the scions on to apple seedlings. By this means Pratt succeeded in obtaining a high frequency of layer changes in 'Giant Spy' and 'Jonathan Sport'. These changes were mostly layer replacements from 2244 to the increasingly diploid 2224 and 2222, but she did obtain a few layer displacements from 2244 to 2444 and 4444 as well. She failed to obtain wholly tetraploid shoots from 'Giant McIntosh' and 'Bates Lobo'. These results reflect the general principle that the replacement of an inner layer, or the corpus, through the duplication of a more external layer is simpler to achieve, and more frequently achieved, than a displacement in the reverse direction. Compared with the spontaneous rate, irradiation greatly increases the frequency of layer changes, but probably has less influence on the direction. Einset (1952) examined the chromosome complements of the roots on self-rooting scions of 'Giant Spy', 2244, and 'Giant Rome', 2224. In the former, the cells were all tetraploid, whereas in the latter they were mostly diploid. Evidently the roots were largely derived from L III and rarely from the more deep-seated L IV tissue. If adventitious shoots were now encouraged to develop in the tetraploid roots, they in turn would be tetraploid. A further assessment of variation in apple cytochimeras was made by Pratt et al., (1967), in which they cut some trees down to stumps, observed the structure of the ensuing suckers, and compared these with the variation of unpruned trees over a wide range of varieties. These included trees with the three main types of apple cytochimera, plus some trees which also had shoots with the structure 2422 (Table 6.2). A wide range of bud variants was found, from which the authors concluded that the periclinal chimeras of apple were inherently plastic, although the degree of plasticity seemed to vary with the chimeral pattern and the variety. As with irradiation treatment, techniques which increased the frequency of bud variants probably did not affect the type of variant; moreover, the variants, however caused, seemed to occur more readily in axillary or rudimentary buds than in fully developed terminal shoot apices.

Haploid-diploid cytochimeras

After X-raying the haploid *Antirrhinum majus* 'Nivea', a variant arose, called 'Wettsteinii', with hair on the leaf petioles and midribs, radially instead of bilaterally symmetrical flowers, and altered anthers and pistils (Melchers and Bergmann, 1958; Melchers, 1960). When it was treated with colchicine to induce chromosome doubling male fertility was restored (Melchers, 1960). Crosses with 'Nivea' produced a wholly 'Nivea' F_1, as did backcrosses between the F_1, and 'Wettsteinii', and crosses with the wild-type also failed to produce any segregating 'Wettsteinii' progeny (Melchers and Labib, 1970). At the time this result was interpreted as possible selection against the 'Wettsteinii' genotype, while the occasional appearance of a pure 'Nivea' shoot was regarded as a back-mutation. To Pohlheim (1978a), however, the breeding results and the bud variation were clear symptoms of a thin-skinned periclinal chimera – a possibility that had previously not been seriously considered. Further support for the chimeral interpretation was demonstrated by Pohlheim when he discovered that the leaf epidermis was haploid ($2n=x=8$) and the mesophyll tissue diploid ($2n=2x=16$). The haploid epidermis had smaller cells and nuclei, and fewer chloroplasts in the guard cells, than the diploid epidermis of pure 'Nivea' and wild-type leaves. Evidently the colchicine treatment had induced diploidy in L II and L III but not L I. The pure 'Nivea' shoots arose by displacement of the haploid L I by the diploid L II and L III:

$$HHD\,(WNN) \quad \text{to} \quad DDD\,(NNN)$$

In addition, some leaves were highly twisted and crumpled, or had crumpled sectors; anatomical sections showed that the affected mesophyll was now haploid, so evidently periclinal divisions in L I had replaced L II:

$$HDD\,(WNN) \quad \text{to} \quad HHD\,(WWN)$$

Pure haploid 'Wettsteinii' shoots were not observed.

Thuja plicata is a conifer with $2n=2x=22$ chromosomes, but Pohlheim discovered one form, *T. plicata* 'Gracilis', that was haploid, $2n=x=11$ (Pohlheim 1968), and another form, *T. plicata* 'Excelsa' that was triploid, $2n=3x=33$ (Pohlheim 1970b). The haploid is particularly interesting, not simply because it is the first discovery of a haploid Gymnosperm, but because it has potential for experimental and practical development.

Pohlheim (1972b) found the haploid especially sensitive to X-irradiation. After a single dose of 0.5 krad there were almost no survivors, whereas Pötsch (1966b) had almost the same lethality only after treating *Abutilon* × *hybridum* with 5.0 krad, and there were comparable differences in magnitude after successive lower doses. One advantage of the haploid is that recessive mutations are immediately manifest, and Pohlheim (1972a) obtained examples of chlorophyll mutations, and growth mutations with needle-like leaves or thread-like, pendulant branches (Pohlheim 1977a). Another advantage of the haploid is that it often reverted spontaneously to the diploid state within the cells of the shoot apex. This diploidization had the effect of creating haploid-diploid periclinal chimeras (Pohlheim, 1971b, 1972a, 1977c, 1980a). The shoot apex of *T. plicata* 'Gracilis' has two temporarily independent layers, but as periclinal divisions frequently occur in L I, periclinal constitutions do not last long. Diploidization in L I created a two-layered chimera with a diploid skin and haploid core, *DH*, and diploidization in L II created the reverse chimera with haploid skin and diploid core, *HD*. The shape of the leaves is dependent on the constitution of the core, and so, as diploid nuclei and cells are larger than haploid ones, the *HD* chimera had large leaves like the normal diploid tree, whereas the *DH* chimera had small leaves like the haploid tree. The chimeral constitution became perceptible after bud variations in which large leaves were borne on shoots with small leaves, and small leaves were borne on shoots with large leaves (Fig. 6.3).

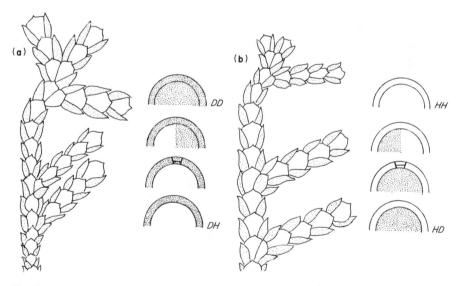

Fig. 6.3   Illustrations of the effect of periclinal divisions in L I of the shoot apex upon the stability of the haploid-diploid cytochimeras of *Thuja plicata* 'Gracilis'. The accompanying diagrams show the changes occurring in the growing points. **(a)** A shoot with small leaves, having L I diploid, L II haploid, changes to one with large leaves – *DH* to *DD*. **(b)** A shoot with large leaves, having L I haploid, L II diploid, changes to one with small leaves – *HD* to *HH*. (Adapted from Pohlheim, 1980a.)

By combining green and white chlorophyll chimeras with haploid-diploid layers, Pohlheim (1977c, 1980a) was able to determine the pattern of growth of the shoots. He found that L I duplication, which was rare in young shoot apices, became more frequent in distant ones. On average a sport occurred once in every ten leaf pairs in *T. plicata* 'Gracilis' compared with once in every 40 leaf pairs in *Juniperus sabina* 'Variegata'. Because of the high rate of L I duplication, the dissociation of the chimeral apex of 'Gracilis' took place before a sufficiently large number of side shoots of a similar structure were produced. The shortage of side shoots with growth potential arose because the frequent periclinal divisions in L I gave rise to white shoots (*WG* to *WW*) which died. Hence, although periclinal apices occurred, they could not be called periclinal chimeras in the strict sense of Winkler, as they could not be guaranteed to maintain the structure through vegetative reproduction (Fig. 6.4).

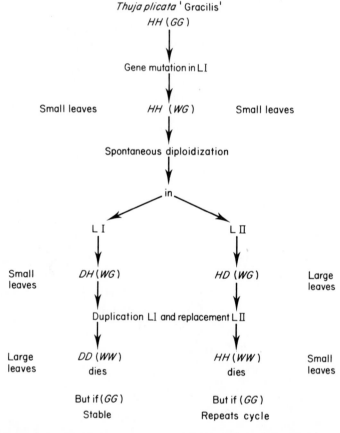

Fig. 6.4 Diagram to explain the causes for the switch in leaf size during shoot growth in the haploid *Thuja plicata* 'Gracilis', together with the effect on white-over-green periclinal apices in this gymnospern. Key: *HH* solid haploid; *DD* solid diploid; *HD* haploid L I, diploid L II; *DH* diploid L I, haploid L II. *GG* solid green; *WW* solid white; *WG* white L I, green L II.

Histogenesis

The wide range of cytochimeras available in the apple, cranberry, peach and thorn apple, and to a lesser extent other species, has rendered them exceptionally valuable for studying the relationship between the tunica and corpus of the apical growing-point and the tissues to which they give rise. The advantage of chimeras compared with wholly diploid plants is that the polyploid cells are cytologically marked by having a larger cell size, larger nuclei, and a higher chromosome complement than the diploid cells (Fig. 6.5). By a suitable choice of chimeras each layer can be specifically marked in turn. Thus in the thorn apple L I was marked with a 422 chimera, L II with a 242 chimera, and L III with a 224 chimera. A second series of cytochimeras had octoploid cell markers, for L I an 822, for L II a 282, and for L III a 228 chimera (Satina et al., 1940). As a check that polyploidy was not itself responsible for modifying development, there were chimeras in

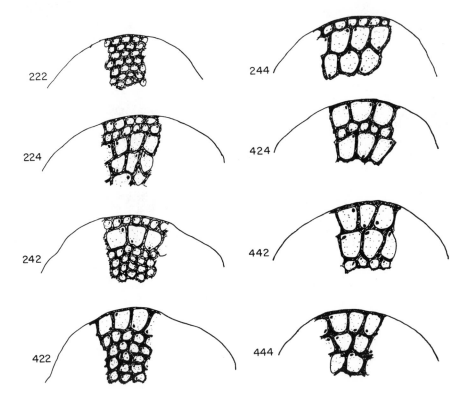

Fig. 6.5 Diagrams of the summits of shoot apices illustrating the eight cytological types observed as a result of the induction of polyploidy following colchicine treatment. Relative to the smaller diploid cells, the size of the larger cells is somewhat exaggerated for tetraploid cells, but quite realistic for octoploid cells.

which diploid cells could be used to mark each layer, for L I 244 or 242, for L II 424 or 224, and for L III 422 and 242 chimeras. Finally, the trichimeras 284 and 428 could be used to mark all three layers differentially at the same time. Rather fewer chimeras were available for cranberry and peach, and there were no octoploid types, but so many chimeras are not essential. The apple chimeras differed from the others owing to the need to consider four layers and not just three, and also by the absence of any chimeras with a tetraploid L I.

The results of these investigations provided very detailed descriptions of the histogenesis of the four species. What they showed was so much variability between them as to make generalizations rather risky, but a few points can be made. The outermost tunica, L I, always gave rise to a single cell thick layer, the epidermis, which covered all organs. In just a few places, cells derived from L I proliferated to develop tissue several cells thick as, for example, the integument around the ovules in all four species (Satina and Blakeslee, 1943; Satina, 1945; Dermen, 1947a; Blaser and Einset, 1950), and the suture of the ovary wall in the peach (Dermen and Stewart, 1973). There is also general agreement that the germ cells, both male and female, develop in sub-epidermal tissue, which in these and most other species is normally derived from L II.

The development of the leaves was variable. The leaf blade was derived entirely from L II in the peach (Dermen, 1953b; Dermen and Stewart, 1973), and mostly from L II in the thorn apple (Satina and Blakeslee, 1941); in the cranberry leaf L II contributed to the marginal mesophyll and the terminal part of the leaf, while L III contributed to the more central and basal areas (Dermen, 1947a). Finally, in the apple L II contributed mostly to the leaf margins, L III to the more central areas (Dermen, 1951b) and, according to Blaser and Einset (1948), L IV contributed to the most central vascular tissue. Dermen (1951b) did not agree with the necessity to distinguish an L IV layer in the development of the leaves and floral organs of the apple.

The vascular tissues usually had the same ploidy as the mesophyll cells through which they ran, so the part of a vein nearest the midrib was often derived from L III and the part nearest the margin from L II. Of course, sometimes, the margin between L II and L III, or L III and L IV, might correspond with the margin between mesophyll and vascular tissue so that in such locations the two types of tissue were derivatives of adjacent apical layers, but such observations should not be extrapolated to the vascular system as a whole.

The floral organs have been intensively investigated and like the leaves there is variation between the species. Apart from the epidermis, the petals and sepals of the thorn apple (Satina and Blakeslee, 1941) were derived largely from L II except at the base, and the petals of the peach were entirely from L II (Dermen and Stewart, 1973); by contrast, the thicker sepals of the peach had more tissue of L III origin than the leaves. The petals and sepals of cranberry (Dermen, 1947a) and apple (Dermen, 1953a) both had marginal tissue derived from L II and central tissues from L III. Similarities and differences of these kinds were also found in respect of other floral parts

– the stigma, style and ovary, and the anthers and their filaments (Satina and Blakeslee, 1943; Satina, 1944).

What is clear from the findings in these and other less intensively examined species is that there is no direct relationship between the apical layers and the tissues to which they give rise. There is, of course, a 'rough-and-ready' relationship in that for all plant parts the order of tissues from the outside to the inside corresponds to the order of the apical layers. But, whereas L I must always participate in organ development, the likelihood of the other layers doing so decreases slightly with L II and quite considerably with L III and L IV. Hence, whether leaf cells develop into spongy or palisade mesophyll, or into vascular tissue, does not depend on the layers from which they originate, but upon which layers are utilized in the plant and the positions the cells find themselves in the differentiating tissues. It is therefore found that comparable cells occupying similar positions in the organs of different species may or may not be derived from equivalent layers in their respective shoot apices. Each species has its own characteristic pattern of development.

7 Variegated-leaf Chimeras

Problems and solutions

There are innumerable variegated-leaf plants grown throughout the world. Their decorative foliage has made them very popular for public parks and botanic gardens, as well as for the home, indoors and outdoors. They are frequently found brightening up the walkways of shopping centres, and adding soothing colour to the foyers of public buildings and office blocks. The multiplication of variegated-leaf plants to supply this demand is a major horticultural industry. The enormous variation in their appearance adds to their overall appeal and makes them of considerable scientific interest.

The development of the chimera concept was seen to provide an acceptable explanation for the structure of graft-hybrids, and for the white-margined leaves of zonal pelargoniums, so it is not surprising that attention was soon paid to many of the other variegated plants in cultivation. Were some of these periclinal chimeras too? Many plants were soon examined; their morphology described; their variations noted; their tissues sectioned; their seedlings scored. The result was overwhelming. An early attempt to bring a little order was made by Küster (1919), who divided variegated plants with white margins into four kinds, which are briefly summarized below.

(i)   The white-margined cultivars of *Abutilon pictum*, *Acer negundo*, *Brassica oleracea*, *Buxus sempervirens*, *Cornus alba*, *Fuchsia magellanica*, *Ligularia tussilaginea*, *Nicotiana gigantea*, *Pelargonium* × *hortorum*, *P. peltatum* and *Solanum dulcamara* among dicots, and *Clivia miniata* and *Dracaena santeri* among monocots, had a white skin that developed all the mesophyll tissue at the leaf margins and formed an unbroken layer, at least one cell thick, right across the leaf blade inside both upper and lower epidermis.

(ii)   *Saxifraga stolonifera* and *Solanum sisymbrifolium* were separated from the first group on account of the heavy peppering of green spots in the white tissues.

(iii)   A variegated *Ligustrum ovalifolium*, and to a greater extent, *Spiraea japonica*, developed highly irregularly. The main shoots of the *Spiraea* were green, but frequently the leaves were sectored with white, or were white-margined, and often wholly white shoots were formed.

(iv)   The white-margined forms of the dicots *Ilex aquifolium* and *Sambucus nigra* and, among monocots, *Agave americana*, *Chlorophytum capense*, and species of *Hosta*, had white mesophyll tissue at the leaf margins, as in the first group, but there was no layer right across the leaf blade covering the green core.

Some species were particularly interesting because of the existence of cultivars in which the arrangement of tissues was the reverse of white-margined forms. Thus there were white-margined, green-centred and green-margined, white-centred forms of *Euonymus japonica* and *Ilex aquifolium*, and of species of *Chlorophytum* and *Hosta lancifolia*. In the case of *Acer negundo* and *Ligustrum ovalifolium*, Küster observed leaves of the alternative types on the same plants. Bateson (1919) too had observed the reversal of the white-over-green leaf to a green-over-white leaf – as if the skin and core had turned inside out – in *Coprosma baueri*, *Euonymus japonica*, *Pelargonium peltatum* (Fig. 5.2) and *P.* × *hortorum* 'Madame Salleron'. Other examples soon followed, and Bateson (1921) was struck by the observation that reversal was always from white-over-green to green-over-white and never in the reverse direction.

Occasionally, there were green flecks on the edges of white-margined plants, and in *Hydrangea macrophylla* (Sabnis, 1923; Chittenden, 1926, 1927) and *Symphoricarpos orbiculatus* (Küster, 1927) green teeth and green lobes, peripheral to the white margin, were very common (Fig. 7.1). In another form of *H. macrophylla* and in *P.* × *hortorum* 'Golden Brilliantissima' the peripheral lobes and teeth were yellow. Many white-margined plants occasionally gave rise to white shoots, and on a few of the white leaves of *Acer negundo*, *Euonymus radicans*, *P.* × *hortorum* 'Madame Salleron' and *Ulmus procera* (Küster, 1919) small green islands, or green sectors, were found on the edges or undersides.

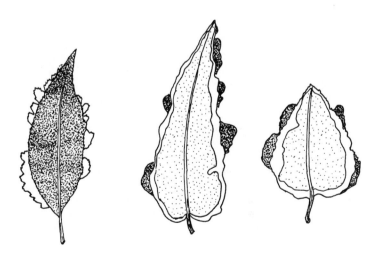

Fig. 7.1 Three examples of leaves from chimeras in which marginal lobes or teeth are derived from localized proliferation of cells derived from L I. **Left** The thin-skinned, *WGG* (*XYZ*) trichimera, *Prunus cerasifera* 'Hessei'. **Centre** The *GWG* sandwich chimera, *Ficus radicans*. **Right** The *GWG* sandwich chimera, *Symphoricarpos orbiculatus*. (Adapted from Bergann and Bergann, 1960; and Pohlheim, 1970a.)

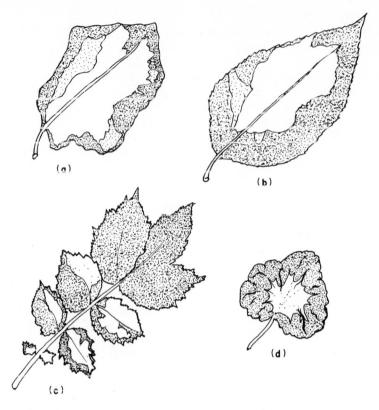

Fig. 7.2 Four examples of leaves from thin-skinned *GWW* chimeras showing an almost continuous green margin developed from L I and surrounding a white core. **(a)** *Aralia pennocki*; **(b)** *Hydrangea macrophylla* 'Nivalis'; **(c)** *Filipendula ulmaria*; **(d)** *Pelargonium* × *hortorum* 'Freak of Nature'. (Adapted from Chittenden, 1927; Dermen, 1960; and Tilney-Bassett, 1963a.)

A very curious looking plant was *P.* × *hortorum* 'Freak of Nature', in which the leaves varied, from those which were almost white, to others, which had large, green, billowing, marginal lobes leaving only a small white centre (Bateson, 1919; Chittenden, 1926) (Fig. 7.2). Rather similar plants were found in *Filipendula ulmaria* (Bateson, 1921) (Fig. 7.2) and *H. macrophylla* 'Nivalis' (Sabnis, 1923; Chittenden, 1926). These plants were quite distinct from the green-over-white plants of the type derived by reversals, in which the white centres were covered by a thin skin of green, and from the green-margined plants with no covering skin over the white centres, but which were of a uniform pattern and regular shape.

Additional examples and further analysis, especially by anatomical studies, soon followed (Funaoka, 1924; Chittenden, 1927; Küster, 1927; Schwarz, 1927; Massey, 1928; Linsbauer, 1931; Sabnis, 1932), but with so many forms of variegated-leaf, and of such varying behaviour, there was some doubt as to whether the chimeral hypothesis was adequate to explain them all. Collins (1922) regarded the white-margined and green-margined forms of *Chlorophytum comosum* and *C. elatum* as having the structure of thin-skinned periclinal chimeras, but he explained the development of these alternative types as the result of an orderly segregation of an inhibitory factor at the somatic meristem. Noack (1922) studied the histological development of the leaf in zonal pelargoniums and concluded that in all dicots, all tissues of the leaf lamina, apart from the epidermis, were products of the division of a single apical sub-epidermal cell. He further assumed that in variegated plants the alternative green and white tissues were the result of somatic segregation from this cell; owing to the action of a labile gene, divisions in one plane produced green cells and in another plane white cells. As this hypothesis meant that white or green-margined plants could not be true chimeras, he called them mantel chimeras (Chittenden, 1927).

There were so many objections to these non-chimeral hypotheses that they were not widely accepted. Nevertheless, they reflected the unease. Many of the plants discussed were regarded not as periclinal, but as sectorial chimeras, or aclinal chimeras (Funaoka, 1924), and Küster (1927), who was reticent about the chimeral hypothesis, felt, in order to explain the appearance of green flecks on the white margins of so many variegated leaves, that it was necessary to assume that green cells regularly arose out of the white cells of organs and organ parts.

Why was it so difficult to fully appreciate the nature of these plants? I believe the solution was far from apparent because the variation between plants arises firstly in differences in the basic structure of periclinal chimeras, and secondly from the many ways in which the expression of these structures is modified during leaf development. Many of the morphological and anatomical studies were trying to explain the second cause of variation with a totally inadequate understanding of the first; only gradually was the situation turned around.

Hanstein's (1868) histogen theory had taught that the outermost layer of the shoot apex – the dermatogen – divided solely by anticlinal divisions to develop into the epidermis covering of all plant surfaces. It was not considered that this specialized tissue might sometimes divide periclinally and so make a contribution to the leaf mesophyll. It was therefore not regarded as having any role to play in the appearance of variegated leaves and so was largely ignored. Only slowly did observations accumulate to show that far from having a neutral role, the behaviour of the epidermis was of the greatest significance, and was the key to a full understanding of the basic structure of the variegated-leaf chimeras. There were several kinds of observations that were important.

Two forms of the monocot *Hosta fortunei* had opposite leaf variegations – white-margined and green-margined. When Chodat (1919) examined the guard cells of the leaf epidermis, he found colourless plastids in the white-margined form, both over the white margin and the green centre, and green plastids in the green-margined form, both over the green margin and the white centre. So, as there was an exact correlation between the colour of the margin and the colour of the plastids in the guard cells all across the leaf, Chodat concluded that the margin was developed from the epidermis. The

Fig. 7.3   *Hosta fortunei* 'Thomas Hogg'. The thin-skinned chimera has a white skin over a green core, *WG(G)*. The marginal tissue is derived from L I, and is very regular.

conclusion was supported by breeding results, firstly in the monocot *Chlorophytum elatum*, for which there was a similar opposite pair of variegated-leaf chimeras (Collins, 1922), and later in variegated forms of *Hosta lancifolia* (Yasui 1929) (Fig. 7.3). In all cases in which the chimeras were used as female parents, the seedlings, which are assumed to be derived from the sub-epidermal layer, corresponded to the central leaf tissue. Hence the marginal tissue was not derived from the sub-epidermal layer, but was of a different layer exterior to it, namely the epidermis. It was therefore evident that these monocots had a two-layered growing-point in which the white-over-green chimeras had the layer structure *WG* and the green-over-white chimera the structure *GW*, or possibly *WGG* and *GWW* respectively. Buder (1928) suggested that the margin developed from the epidermis in *Agave*, *Chlorophytum* and *Sanseviera*, and Imai (1935b) argued that the white stripes often seen in the central green areas of white-margined forms of *Ananas*, *Arrhenatherum*, *Arundo*, *Carex*, *Miscanthus*, *Phalaris*, and *Polygonatum* were also derived from L I.

The importance of the epidermis in the structural analysis of dicots was not appreciated so quickly. Instead of a positive correlation between the colour of the leaf margin and the colour of the plastids in the guard cells, Correns (1919) found a negative one. White-margined forms of *Arabis albida* and *Mesembryanthemum cordifolium* had green plastids in their guard cells, and white shoots from *A. albida* had occasional green flecks on the surface of leaves (Fig. 7.4). Correns recognized that the epidermis of the white-margined forms was normal, but this did not lead him to

Fig. 7.4 *Arabis albida* 'Variegata'. The *GWG* sandwich chimera is rather unstable and produces frequent white shoots with a genetically green epidermis.

appreciate its significance even though there were no green flecks on the white seedlings obtained from crosses in which *A. albida, M. cordifolium* and white-margined forms of *Aubrieta deltoidea* var. *graeca* and *Arabis purpurea* were used as the female parent. Two kinds of white shoot were described by Kümmler (1922) as bud variations from *P.* × *hortorum* 'Madame Salleron'. On the surface of one, there were green flecks and green plastids in the guard cells, on the other, there were no green flecks and the plastids were colourless. The same two kinds of white shoot were observed by Chittenden (1926) in bud variations of *P.* × *hortorum* 'Freak of Nature'. Within a few years green chloroplasts in guard cells and green flecks on white leaves or on white edges of chimeral leaves and stipules were observed in over half a dozen other white-margined chimeras (Kümmler, 1922; Rischkow, 1927; Chataway and Snow, 1929). Surprisingly, in spite of the accumulating evidence for a clear association between green chloroplasts in the guard cells and green flecks on the surface of white-margined leaves, the investigators failed to suggest that the white skin of many chimeras might have a genetically green epidermis. On the contrary, in addition to Küster's (1927) vague explanation, Imai (1928, 1936a) considered that the plastids were automutable to green in flecked shoots and constant in non-flecked ones, whereas Neilson-Jones (1934) believed that flecks arose from isolated islands of green tissue scattered throughout the white growing-point. Yet, Neilson-Jones took a different view of the green lobes and green teeth on the leaf margins of *H. macrophylla*. He suggested that the growing-point possessed a green core surrounded by a two-layered skin composed of an inner layer of colourless tissue and an outer layer of green tissue (Fig. 7.5). The yellow lobed forms of *H. macrophylla* and *P. × hortorum* 'Golden Brilliantissima' were similar. Neilson-Jones also argued that since there was no green skin over the central area of the leaves of *H. macrophylla* 'Nivalis', the skin was no more than one cell layer thick, it comprised the epidermis alone, and it produced the green leaf border through periclinal divisions; *Filipendula ulmaria* (Fig. 7.6) and *P. × hortorum* 'Freak of Nature' were similar. Chittenden (1927) had briefly considered this explanation too, but felt he lacked the proof. Hence, twenty five years after Baur first recognized the periclinal chimera, Neilson-Jones had realized that the epidermis was of real significance in chimera structures, and was not always the same as the sub-epidermal layer. This followed the earlier recognition of the importance of the epidermis in graft chimeras, and Imai's (1934, 1935a) recent explanations of flower chimeras. Soon afterwards, Rischkow (1936) realized that a white-over-green cultivar of *Petunia hybrida* had a two-layered skin consisting of a genetically green epidermis and white sub-epidermis. Renner (1936a) likewise concluded that the example of *P. × hortorum* 'Madame Salleron' was a question of a *GWG* mesochimera with a potentially green epidermis, colourless sub-epidermis and green core. He was thus able to explain that the white shoots often possessed a normal epidermis, *GWW*, so that, whenever they produced mesophyll tissue a green fleck appeared. Moreover, Renner pointed out that the often reported reversal could be explained on the assumption of a sandwich structure. Precisely to what Renner was referring was fully explained by Dermen (1950), who showed that reversal was caused by periclinal divisions in the epidermal layer of the growing-point replacing the inner layers – literally pushing them down the stem. In this way a *GWG* sandwich, with white skin and green core, became a *GGW* chimera with green skin and white core (Fig. 5.2). Earlier Bateson (1924) had observed green shoots from white-margined *P. × hortorum* 'Flower of Spring' and 'Caroline Schmidt' developing white flecks and sometimes reconstituting a new white margin. Renner explained these changes through a loss of the original green epidermis, followed by a shift of the original white sub-epidermal layer into the vacant epidermal position, *GWG* to *WGG*. Subsequently, periclinal division of the white epidermis produced a white fleck on the green margin, or even a whole new white margin, *WGG* to *WWG* (Fig. 5.1).

The question of the types of chimera structure had arisen in connection with graft chimeras and flower chimeras, and Imai had shown that with two genotypes there were six possible structural arrangements of three-layered chimeras together with the two pure types. Hence, in respect of green and white variegated-leaf chimeras, there were two types of two-layered chimeras – *GW* and *WG* – and six types of three-layered chimeras – thin-skinned, *GWW* and *WGG*, thick-skinned, *GGW* and *WWG*, and sandwich, *GWG* and *WGW*. Many chimeras are readily placed into one of these types because the pattern of growth in the leaf clearly reflects the structure in the growing-point. Other chimeras are less easy to classify owing to a more complex relationship between leaf development and the tunica corpus

Fig. 7.5   *Hydrangea macrophylla* 'Variegata'. A *GWG* sandwich chimera. The marginal teeth arise as a result of localized periclinal divisions in the genetically green epidermis during leaf development. They are clearly separated from the green core – L II – by a thin layer of white tissue of L II origin. (From Kirk and Tilney-Bassett, 1978.)

Fig. 7.6   *Filipendula ulmaria* 'Aurea'. The thin-skinned chimera has a green skin over a white core, *GW(W)*. The growth of the white core is so inhibited that all the green tissue is derived from extensive division in L I.

Fig. 7.7 Examples of two highly decorative, chimeral grasses. **Above**   *Glyceria aquatica* 'Variegata'. **Below**   *Arrhenatherum elatius* 'Variegatum'.

structure. Moreover, there is often no correlation between potential chimeral structures and major taxonomic grouping.

According to Rösler (1928) the leaves of oat, rye and wheat are developed entirely from the epidermis layer so periclinal leaves are not possible; *Panicum palmifolium* is similar. A periclinal structure is also impossible in

Saccharum officinarum which has no tunica. A single layered tunica occurs in *Oplismenus imbecilis* and the indian millet and, as the corpus contributes to the leaf, chimeras are possible. Still other grasses – *Bambusa verticillata, Dactylis glomerata, Glyceria maxima, Phalaris arundinacea, Phragmites communis* and *Zizania miliacea* – have a two-layered tunica and corpus (Thielke, 1951; Neilson-Jones, 1969) and highly decorative variegated-leaf chimeras of these species are commonly grown in garden borders and the edges of pools. Examples of similarly variegated species are shown in Fig. 7.7. Many other monocots are variable too and include examples based on one or two tunica layers plus a corpus, and Pohlheim (1982a) interprets the structure of *Dracaenas* on the assumption of three tunicas and a corpus. As the dicots exhibit a similar range to the monocots, there is little virtue in trying to distinguish between them; highly dissimilar leaf shapes may develop from similarly structured growing-points.

The complexity in chimera structures arises partly through the different ways in which leaves grow, but mostly through the wide variations that occur between species in the relative contributions of the layers to leaf growth. In addition, whereas some chimeras are highly stable with a fixed number of layers contributing to leaf growth, others are far less stable owing to considerable flexibility in the relationship. For example, *Spiraea japonica* 'Anthony Waterer' normally produces green leaves with no visible sign of the genetically white epidermis; its structure is apparently L I white, L II green (Fig. 7.8). Yet, not infrequently, it produces wholly white leaves and

Fig. 7.8 Left *Spiraea japonica* 'Anthony Waterer'. An eversporting, thin-skinned chimera with apparently green shoots, *WGG*, white-over-green shoots, WWG and pure white shoots, *WWW*. (From Tilney-Bassett (1963a). With the permission of the editors of *Heredity*.) **Right** *Spiraea japonica* 'Gold Flame'. An eversporting, thin-skinned chimera with apparently golden-yellow shoots having a genetically green epidermis, *GYY*, and frequently sporting yellow, green mericlinal and pure green leaves.

shoots in which the chimera structure has broken down, but it also produces, for a brief period, shoots in which the leaves have a clear white margin and a green centre; its structure is now L I white, L II white, L III green. Hence, at different places within the same bush, the shrub behaves as not layered, as single layered, and as double layered plus corpus. It is therefore only as a matter of convenience and in recognition of its potential that I would classify it as three-layered, *WGG*. A similar problem arises in zonal pelargoniums. Many of the white-over-green cultivars have the *GWG* sandwich structure with a uniform white margin and green centre indicative of a three-layered growing-point, but sometimes a bud variation occurs in which the white margin is much wider while a small green centre still remains. Temporarily the growing-point is four-layered with the structure *GWWG* (Stewart *et al.*, 1974). Yet in most descriptions of *Pelargonium* chimeras it is more convenient, and simpler, to think in terms of the more stable three-layered structure. After all, the *GWWG* chimera has so little green tissue and is so unstable that it would never become a useful cultivar, and it hardly matters if the original chimeras might also have periods of being *GWGG* rather than *GWG*; one cannot tell if and when these periods occur, and it makes no difference to the kinds of bud variations to which the sandwich chimera can give rise. A more widespread problem exists when classifying thin-skinned chimeras – are the two kinds two-layered, *GW* and *WG*, or three-layered, *GWW* and *WGG*? Whether two- or three-layered, analysis of seedlings and root cuttings has the same expectation. It is only if the thin-skinned chimeras produced thick-skinned bud variants – *GWW* to *GGW* and *WGG* to *WWG* – or if another cultivar of the same species is a sandwich chimera, *GWG*,can we be reasonably sure that we are dealing with a three-layered structure. It is, therefore, only by dint of their similarity to other species, and for convenience, rather than by direct proof, that Stewart and Dermen (1979) chose to regard many monocots as three-layered rather than two-layered, and also to avoid inferring a difference between monocots and dicots that is more imaginery than real. Once these difficulties are recognized, it becomes more instructive to consider the variation in behaviour of thin-skinned chimeras.

Thin-skinned chimeras, *GWW* and *WGG*

Besides a direct origin through mutation and sorting-out, thin-skinned chimeras can arise by bud variation from thick-skinned chimeras to give apparently green or white shoots:

L I displacement:	*WWG* to *WGG* green shoot
L I displacement:	*GGW* to *GWW* white shoot

or from *GWG* sandwich chimeras:

L I displacement:	*GWG* to *WGG* green shoot
L III replacement:	*GWG* to *GWW* white shoot

The epidermis of these green and white shoots is often stable and the chimera structure remains invisible. The chimera is recognized by the colour of the plastids in the guard cells, or by the occurrence of a rare white or green fleck or sector respectively on a leaf edge. I have observed both green and white chimeral shoots in bud variations from *P.* × *hortorum* 'Flower of Spring', and white shoots with occasional green flecks in a wide range of species (Tilney-Bassett, 1963a), and other authors have reported similar findings. A chimera in peach (Dermen, 1960) appears to be of this kind.

Much more striking are thin-skinned chimeras in which L I frequently contributes highly irregularly to the mesophyll at the edges of the leaf, especially in the teeth and lobes, forming a discontinuous, broken band (Fig. 7.1). A good example is *Prunus cerasifera* 'Hessei' (Küster, 1934; Renner and Voss, 1942). After further careful investigation, Pohlheim (1970a, 1983) has discovered that the tree is a trichimera with the three layers (*X*, *Y* and *Z*) characterized as follows:

L I	White, cell growth normal	(*X*)
L II	Green, cell growth inhibited	(*Y*)
L III	Green, cell growth normal	(*Z*)

He explained the excessive division of L I to give rise to chunks of white mesophyll tissue in the white teeth as a direct response to the inhibited growth of the underlying L II layer. The evidence for this idea was provided whenever a bud variant occurred in which the *Y* layer, was lost and the *Z* layer duplicated. The resulting chimera, *XZZ*, was still *WGG* with respect to the plastid genotypes, but without the growth inhibiting *Y* layer the leaves were green and of regular shape and the now rare, small, white teeth no longer distorted the edges. An unusual form of *Quercus robur* (Küster, 1929) may have been similar. Thin-skinned chimeras with irregular green teeth and lobes were found in *Aesculus* × *carnea* and *A. hippocastanum* (Küster, 1929; Bergann and Bergann, 1982) and in *Aralia pennocki* (Dermen, 1960).

Many mutant plastids have little or no effect on cell growth, but in some mutants plastid degeneration starts before the normal end of cell division, which causes pure mutant cells to stop dividing leading to various degrees of cell and leaf distortion. Early plastid inhibition may lead to the failure of one, two or more of the last successive cell divisions resulting in severe loss of cells and the narrowing of variegated leaves, asymmetrical development, buckling, and general deformation (Kirk and Tilney-Bassett, 1978). In *Prunus cerasifera* 'Hessei' it is likely that the growth inhibition has been caused by a nuclear mutation as the plastids are still green, but when the growth inhibited layer corresponds with the mutant layer the two phenomena are probably either pleiotropic effects of the same mutation or two mutants in one layer.

A similar chimera to 'Hessei', but with a more extensive, very irregular, continuous green skin around a very much reduced white core, is *Euphorbia pulcherrima* 'Weisskern Rosa', a derivative of 'Eckes Rosa' (Bergann and Bergann, 1960; Bergann, 1961, 1962c, 1967a, 1967b), which is again a trichimera:

L I	Green, anthocyanin defective, cell growth normal	(X)
L II	White, anthocyanin intact, cell growth inhibited	(Y)
L III	White, anthocyanin intact, cell growth normal	(Z)

In this case Pohlheim (1983) found the chimera unstable, and wholly green shoots were often produced. Evidently, cell division by the sub-epidermal layer was so inhibited that L I was encouraged into excessive periclinal divisions even in the growing-point. The failure of adequate cell division in the inhibited layer was seen anatomically by the shortage of cells, while the spectacular effect of L II on L I was proved by the normal stability of green L I in white shoots of the chimera XZZ in which layer Y was lost and replaced by a duplicated layer Z.

After many observations and attempts to explain the structure of *P.* × *hortorum* 'Freak of Nature' (Bateson, 1916; Chittenden, 1926, 1927; Noack, 1930; Neilson-Jones, 1934; Imai, 1936b; Ufer, 1936; Renner, 1936a, 1936b; Thielke, 1948; Bergann, 1962a; Tilney-Bassett, 1963a), Pohlheim (1973b, 1977b, 1983) has convincingly proven that it too is a trichimera:

L I	Green, cell growth normal	(X)
L II	White, cell growth inhibited	(Y)
L III	White, cell growth normal	(Z)

The irregular, monstrous, billowing green margin of the leaves of this famous cultivar is again entirely due to the extensive periclinal divisions in the cells of L I producing all the green mesophyll, induced to divide by the stunted growth of the underlying white layer. Pohlheim has carefully analysed many of the derivatives of 'Freak of Nature', which are fully supportive of his interpretation. Other well known GWW chimeras with expansive broad green margins and narrow white centres are *Hydrangea macrophylla* 'Nivalis' (Sabnis, 1923; Chittenden, 1926, 1927; Küster, 1927; Krenke, 1933; Neilson-Jones, 1934; Renner, 1936a; Tilney-Bassett, 1963a) and *Filipendula ulmaria* (Bateson, 1921; Chittenden, 1927; Neilson-Jones, 1934; Imai, 1935b; Thielke, 1948; Tilney-Bassett, 1963a) (Fig. 7.2). In view of Pohlheim's convincing interpretation for similar chimeras, it would not be surprising if these also carried an inhibiting sub-epidermal layer beneath a normal epidermal layer.

Another group of chimeras are eversporting. The leaves of *Spiraea japonica* 'Anthony Waterer' normally appear fully green with no sign of a chimera structure. Yet, quite frequently, bent leaves are produced with a white sector running from base to tip, often from margin to midrib, or a whole leaf is white, or there is a white margin all round, and wholly white shoots are common (Küster, 1927; Imai, 1935b; Renner, 1936a, 1936b; Bergann, 1962a; Tilney-Bassett, 1963a; Pohlheim, 1971f). Evidently, L I is rather unstable and frequently divides in all planes to completely replace the green L II and L III in the development of large sections of mesophyll. Fortunately, the shrub does not become fully white or it would die. The white shoots cannot support their own indefinite growth and gradually come to a halt. Many axillary buds, from below the regions where the growing-points had become fully white, still retain the chimeral structure and these provide a steady supply of new, eversporting green shoots, always covered by the white epidermis. The variegated leaves of *Mentha arvensis* 'Variegata' (Pohlheim, 1971f) originate in a similar fashion, and probably the white sectoring observed in *M. rotundifolia* 'Variegata' (Sabnis, 1932) and in *Salvia officinalis* 'Tricolor' (Tilney-Bassett, 1963a) is also of epidermal origin (Fig. 7.9).

Fig. 7.9 *Salvia officinalis* 'Tricolor'. A thin-skinned chimera having apparently green leaves covered by a genetically white epidermis, *WG(G)*, which not infrequently gives rise to wholly white sectors.

Fig. 7.10 *Daphne × burkwoodii* 'Carol Mackie'. The presumed, thin-skinned chimera, *WGG*, with a narrow white margin edging the leaves, is very stable with no sign of any bud variation in the growing-point, and little developmental variation in the leaves.

A further group of thin-skinned chimeras regularly have a continuous uniform band all round the edge of the leaves. This is little more than a white fringe in *Daphne odora* and *Fragaria chiloensis* (Imai, 1935b) (Fig. 7.10), but is rather broader in *Veronica gentianoides* (Correns, 1920, 1928; Chittenden, 1927; Massey, 1928; Renner, 1936a, 1936b; Renner and Voss, 1942; Tilney-Bassett, 1963a). The white margin appears superficially similar to the white margin of white-over-green sandwich chimeras, but unlike these there is no continuous layer of white cells running all across the leaf just inside the upper and lower epidermis. Moreover, selfed seedlings from *V. gentianoides* (Tilney-Bassett, unpublished) are all green indicating that the green core is derived from L II, and the outer white skin from L I. As there is no distortion of the marginal tissue, and the chimera is extremely stable, it is probable that the regular periclinal division of L I in the leaf primordia, absent in the growing-point, reflects the normal development in this plant. Among dicots a uniform white margin derived from L I is rather uncommon, and likewise a uniform green margin, for which *Sedum albo-roseum* 'Variegatum' is a possible example (Fig. 7.11). Both types seem to be more frequent among monocots. There are both white-over-green, *WG(G)*, amd green-over-white, *GW(W)*, cultivars of *Chlorophytum capense*, *C. comosum* and *C. elatum* (Küster, 1919; Collins, 1922; Correns, 1922; Chittenden, 1927; Massey, 1928; Imai, 1935b; Renner, 1936a, 1936b; Thielke, 1948; Stewart and Dermen, 1979) and similarly both types are well represented among species of the very popular *Hostas* including *Hosta decorata*, *H. fortunei*, *H. ovata*, and *H. undulata* (Chodat, 1919; Küster,

Fig. 7.11 *Sedum albo-roseum* 'Variegatum'. The green-over-white chimera is thought to have a thin green skin, *GWW*, and is stable in development.

1919; Kümmler, 1922; Yasui, 1929; Neilson-Jones, 1934; Imai, 1935b; Renner, 1936a; Thielke, 1948; Stewart and Dermen, 1979). The origin of the wide margin, whether white or green, from L I, and the core from L II, has been well proven by observation, and by anatomical and genetical investigations, but it is unclear whether or not there is any regular contribution from a third layer. There appear to be many similar species, although less well tested, such as the white-margined *Agave americana* (Trelease, 1908; Küster, 1919, 1927; Imai, 1935b), *Ananas comosus* (Imai, 1935b), *Arrhenatherum elatius* (Imai, 1935b), *Arundo donax* (Sabnis, 1932), *Carex morrowi* (Imai, 1935b), *Dendrobium monile* (Imai, 1935b), *Sansevieria trifasciata* 'Laurenti' (Küster, 1927; Thielke, 1948; Dermen, 1960; Stewart and Dermen, 1979) and *Yucca aloifolia* (Trelease, 1908; Imai, 1935b). Comparable green-margined cultivars are less common but a good example is *Agave americana* 'Mediovariegata' (Renner, 1936a) and another *Carex riparia* 'Variegata' (Tilney-Bassett, unpublished). In some of these chimeras L I produces only the leaf margins, but in others periclinal divisions may also create stripes of green or white mesophyll tissue, derived from the green or white epidermis, growing down into the core zone.

Thick-skinned chimeras, *GGW* and *WWG*

Thick-skinned chimeras may arise either directly following mutation and sorting-out of plastids, or as bud variations from thin-skinned chimeras:

L II replacement:	*GWW* to *GGW*	green-over-white shoot
L II replacement:	*WGG* to *WWG*	white-over-green shoot

or from *GWG* sandwich chimeras:

L II replacement:	*GWG* to *GGW*	green-over-white shoot
L I displacement:	*GWG* to *WWG*	white-over-green shoot

The green-over-white shoots are usually far less striking than the white-over-green, and so the chimeras are not so highly valued for garden display (Fig. 7.12). It is often difficult to see much contrast in the colours of the leaf in the green-over-white chimera because the weak marking is often a pale, narrow, indistinctly bordered, middle field, owing to the covering over of the white core, above and below, by green tissue. A white-over-green chimera in *Hypericum tetrapterum* had a very wide border and narrow green centre, so in the reverse chimera the small white core would probably be completely obscured (Noack, 1932). In *Ligustrum ovalifolium* (Küster, 1919) the border is not so much white as pale green, which would create little contrast in the core position of the reverse chimera. These chimeras often arise by the reversal of a white-over-green sandwich chimera (Fig. 7.12), but mostly they have too little commercial value to warrant keeping. Occasionally the marking is sufficiently strong to be attractive; the shining leaves of the green-over-white *Vinca major* 'Oxoniensis' make an effective foil to the white-over-green leaves of *Vinca major* 'Variegata', *GWG*, from which it arose (Tilney-Bassett, 1963a).

Fig. 7.12 Two contrasting forms of the greater periwinkle. **Above** *Vinca major* 'Elegantissima', *GWG*. **Below**   *Vinca major* 'Oxoniensis', *GGW*, which was derived from the *GWG* sandwich chimera by reversal.

The white-over-green chimeras are attractive plants with conspicuous leaf borders, but mostly they turn out to have a green epidermis, and ones with both L I and L II white are rare. They are recognized because of the colourless plastids in the guard cells, and by the absence of occasional green flecks on the edges of the white margins, which are so typical of the sandwich chimeras. Kümmler (1922) recognized *Pelargonium* × *hortorum* 'Mrs Parker' and 'Mr Langguth' as thick-skinned chimeras, and Renner and Mitarbeiter (1952) mention *Ficus elastica*. Duckett and Toth (1977) found giant aggregations of mitochondria in leaf cells from the mutant layer of *Ficus*, whereas the normal green mesophyll cells had the usual simple mitochondria. Another thick-skinned *Pelargonium* is 'Chelsea Gem', *WWG*, which was used by Jamieson and Willmer (1984) to study the functioning of stomata lacking guard cell chloroplasts. They found that the stomata responded more or less normally to light and carbon dioxide as gauged by direct measurements of stomatal aperture and by transpirational water loss studies. The stomata with mutant plastids were more reluctant to open than those with normal chloroplasts, but they still accumulated K^+ ions in the guard cells indicating that chloroplasts were not essential for the normal functioning of the stomata, and that the energy source for driving stomatal movement could come from sources other than photophosphorylation.

An unusual chimera is *Sambucus nigra* 'Albovariegata'. The leaves have a rather narrow white margin, and the white tissue does not form a continuous layer right across the leaf just inside the upper and lower epidermis. Hence the white margin is actually derived by periclinal divisions in the epidermis layer of the leaf primordia. Nevertheless, it is not a thin-skinned chimera. Anatomical investigations have revealed the existence of a few white cells squeezed between the epidermal layer and the green mesophyll layer across the centre of the leaf, enabling Pohlheim (1982b) to interpret the tree as another trichimera:

L I	White, cell growth normal	(*X*)
L II	White, cell growth inhibited	(*Y*)
L III	Green, cell growth normal	(*Z*)

He suggests that the few white cells across the centre of the leaf are the remnants of a much reduced L II owing to the severely restricted growth of these cells, and it is owing to this inhibition that cells of L I have been stimulated to divide periclinally just as in some of the thin-skinned chimeras previously discussed.

An interesting green-margined chimera in which the white core does show up quite well is the *Pelargonium* 'A Happy Thought'. Unlike most variegated-leaf chimeras tested (Kirk and Tilney-Bassett, 1978), this cultivar is a nuclear gene differential chimera. The germ layer is heterozygous and the pale core homozygous recessive for a nuclear mutation (Tilney-Bassett, 1963b). The pigments within the sub-epidermal layer of mesophyll tissue are much reduced compared with the concentration found in the typical green-over-white plastid gene differential chimera, which suggests that the action of the mutant gene upon the plastids

Fig. 7.13   Three leaves illustrating thick-skinned chimeras, *GGW*, in which the white core is not masked by a green skin. **Left**   *Euonymus japonica* 'Mediopictus'. **Centre**   *Elaeagnus pungens* 'Aureo-variegata'. **Right**   *Pelargonium* × *hortorum* 'A Happy Thought'. (From Tilney-Bassett (1963a). With the permission of the editors of *Heredity*.)

within the core cells spreads into the immediately adjacent skin cells. The result is that the core shows through as a broad pale zone contrasting strongly with the dark green border (Fig. 7.13).

There are several chimeras which have a green margin and unmasked white core, apart from rare leaves. These include *Aucuba japonica* (Imai, 1935b), *Coprosma baueri* 'Picturata' (Bateson, 1919, 1921; Küster, 1926, 1927; Sabnis, 1932; Tilney-Bassett, 1963a), *Elaeagnus pungens* 'Aureo-variegata' (Küster, 1927; Imai, 1935b), *Euonymus japonica* 'Mediopictus' (Bateson, 1919, 1921; Küster, 1919, 1926, 1927; Imai, 1935b; Dermen, 1960), several cultivars of *Hedera helix*, *Hoya carnosa* 'Variegata', several cultivars of *Ilex aquifolium* (Küster, 1919, 1926, 1927; Bergann, 1962a), *Ilex myrtifolia*, (Fig. 7.14) and *Sedum sieboldii* 'Medio-variegatum' (Funaoka, 1924; Chittenden, 1927; Imai, 1935b). It is not known if these are all of similar structure, but for most of them there are complementary white-over-green cultivars, and in several the green-margined cultivars have been observed to arise by reversal from the white-margined types. As some of these chimeras had green margins, but no green skin across the white centre beneath the upper epidermis, they were initially regarded as peculiar (Chittenden, 1927) or as aclinal (Küster, 1927, 1937) chimeras. Rischkow (1936) repudiated this explanation for *Elaeagnus pungens* when he realized that the margin was developed from the sub-epidermal layer and not from L I in which there was no evidence for the required periclinal divisions. The origin by reversal is proof enough of the *GGW* structure, so why do the leaves lack the usual covering of green palisade mesophyll cells inside the upper epidermis? Imai (1934, 1935b) explained the transparency of the leaf centre on the assumption that the mutant plastids in these core cells secreted a toxic substance which bleached out the green colour of the overlying cells; much later doubts about this explanation were expressed (Tilney-Bassett, 1963a) as it assumed a property for mutant plastids for which there was no

Fig. 7.14 *Ilex myrtifolia* 'Aurea'. The presumed, thick-skinned green-over-white chimera, *GGW*, has an unmasked core owing to the colourless hypodermal layer present in holly leaves. The size and shape of the green edge is quite variable.

independent evidence. Recently Bergann and Bergann (1983b) have reinvestigated this problem by comparing the anatomical structures of the white-over-green and green-over-white cultivars of *Coprosma baueri, Elaeagnus pungens, Hoya carnosa* and *Ilex aquifolium*, plus the white-over-green form of *Nerium oleander*. In each species, they found that leaf development differed from the normal type of periclinal chimera by the presence of a water tissue underlying the epidermis. This hypodermis layer, developed from L II, consisted of one or two roughly isodiametrical cells devoid of chloroplasts. The elongated palisade cells were not absent, but developed beneath the hypodermis either from L II in the region of the green margin or from L III in the centre of the leaf. It follows that as the layer of cells which ran right across the leaf belonged to the colourless hypodermis, the underlying colourless core tissue was not masked and showed through clearly (Fig. 7.15).

According to Thielke (1954), *Commelina benghalensis* 'Foliis variegatus' had pale green leaves with white margins and white stripes running into the tips. The thick-skinned chimera had a white skin and green core, *WWG*. The pale green parts of the leaf had chlorophyll in the central mesophyll and practically none in the outer mesophyll and epidermis. At the leaf margin the wholly white tissue was caused by the normal absence of corpus tissue, whereas the white stripes in the central green zone were caused by the abnormal absence of corpus derivatives as these were often replaced in the leaf primordium below the growing-point by cells of the inner tunica layer resulting in alternating green and white stripes (Clowes, 1957; Neilson-

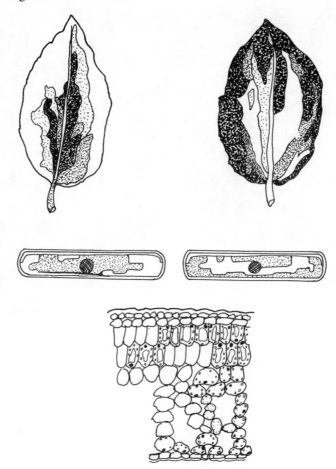

Fig. 7.15 Illustrations of the variegated-leaf chimeras of *Elaeagnus pungens*. **Upper left** The *GWG* sandwich chimera. **Upper right** The *GGW* reversal, thick-skinned chimera. **Centre** Diagrams of transverse sections taken through the middle of the leaves above. **Below** Section through a leaf at the transition between a white and green zone of the thick-skinned chimera. Notice particularly how the existence of a colourless hypodermal layer of water tissue beneath the upper epidermis allows the core tissue, from L III, frequent opportunies to perforate the skin tissue of opposite colour from L II. (Adapted from Bergann and Bergann, 1983b.)

Jones, 1969). *Tradescantia fluminensis* 'Albostriata' was also an unusual chimera. At the shoot apex there was no continuous tunica, but there were some colourless outer layers which in some places were displaced by the outer cells of the green corpus. Consequently, the leaf primordium was formed partly from cells derived from the outer white layer and partly from the green inner tissues, a mixture which was made apparent in the green and white stripes of the leaf. Essentially the plant is a periclinal chimera that is secondarily sectored (Clowes, 1957; Neilson-Jones, 1969).

Sandwich chimeras, *GWG* and *WGW*

Most chimeras originate with a spontaneous mutation occurring within a single cell of the growing-point, followed by the further divisions of the mutant cell and its daughters until their successors occupy a complete layer. The chimeras *WGG*, *GWG*, and *GGW* are likely to arise equally frequently, as each requires only a single mutation in one of three layers. The chimeras *WWG*, *GWW*, and *WGW* are likely to arise infrequently by two separate mutations as the low probability of occurrence is the product of the rerquency in each of two layers. The fact that the first two, *WWG* and *GWW*, are not uncommon is because they often arise indirectly as bud variants of chimeras with a single mutant layer. This second pathway is closed to the *WGW* sandwich chimera and so it is rarely found. In zonal pelargonium it may arise by sorting-out from variegated embryos or from a rare breakdown and rearrangement of the layers (Bergann and Bergann, 1962). Of the three chimeras arising direct from mutation, the *WGG* type is often overlooked because the white L I contributes only to the invisible epidermis. Similarly, the *GGW* type is often overlooked because the white L III tissue contributes only to the hidden core of the leaf. Even when not overlooked, these two types are often, but not always, unattractive ornamental plants and so are then not propagated as useful cultivars. By contrast, the *GWG* chimeras usually develop as highly attractive plants with white-margined leaves, easily spotted and assiduously propagated, making them the most prolific of all chimeral types.

Fig. 7.16 *Abutilon pictum* 'Variegata'. A *GWG* sandwich chimera. In the upper half of one side of the leaf there is a massive green lobe derived from extensive cell division of the genetically green epidermis.

The *GWG* sandwich chimera has been reported in many species (Table 7.1) and is often easy to recognize. The outermost layer, L I, is recognized by the presence of green chloroplasts in the guard cells and, when present, in other epidermal cells of the leaf. It is also identified by the rare to frequent occurrence of green flecks on the edges of the white-margined leaves, stipules, bracts and sepals, and always in contact with the epidermis (Fig. 7.16). The central layer, L II, is recognized by its position and colour, and by the seedlings which are invariably yellow or white like the colour of the leaf margins. The inner layer, L III, is recognized by its position and colour, and by root propagation in which adventitious buds develop from the green core. These chimeras are noted too for the wide range of bud variations to which they can give rise, which includes every type of chimera structure, except the reciprocal *WGW* type, plus the solid green and white derivatives. The reversal of white-over-green to green-over-white, owing to the change *GWG* to *GGW*, is particularly striking (Fig. 5.2). Not all bud variants are equally frequent and in a survey of over two dozen sandwich chimeras, Tilney-Bassett (1963a) found that the most common bud variations were those involving duplication or triplication of either L I or L II and consequent replacement of the adjacent inner layers:

replacement L II/L III: *GWG* to *GGW* to *GGG*
replacement L III: *GWG* to *GWW*

Only in *P.* × *hortorum* 'Flower of Spring' was L I displaced by L II in the change *GWG* to *WGG*; a similar change was recorded in an unnamed mutant of *Pelargonium* by Stewart *et al.* (1974).

Like the thin-skinned *GWW* chimeras, the *GWG* sandwich chimera varies greatly between species in the stability of L I. In some species L I is extraordinarily stable and even tiny flecks of green mesophyll on the edges of the white margins are extremely rare. In other species such flecks, indicative of periclinal divisions in L I, are not uncommon. In yet other species, L I divides periclinally so frequently in leaf development that there is a broken green edge to the white margin, as in *Symphoricarpos orbiculatus*, or the teeth are regularly green as in *Hydrangea macrophylla* 'Variegata' (Fig. 7.5), or yellow in the *YWG* cultivars of *H. macrophylla* and the *Pelargonium* 'Golden Brilliantissima'. Whereas most sandwich chimeras have been only briefly reported, a few have been the subject of more exhaustive investigations.

An interesting case is the poinsettia, *Euphorbia pulcherrima* 'Weißrand Rosa', which is a trichimera with the following structure:

L I   Green, anthocyanin synthesis blocked   (*A*)
L II   White, anthocyanin synthesis normal   (*B*)
L III   Green, anthocyanin synthesis normal   (*C*)

Bergann (1961) accounted for the origin of this plant through two successive, independent mutations. The first mutation blocked anthocyanin synthesis in the epidermal layer of the red flowered 'Imperator' (*CCC*) to give 'Eckes Rosa' (*ACC*). The second mutation blocked chloroplast development in the cells of the sub-epidermal layer to produce the

Table 7.1 Examples of dicots having the *GWG* sandwich structure.

Species	Cultivar	Investigators
Abutilon × *hybridum*	*Andenken an Bonn*	Lindemuth, 1907; Küster, 1919, 1927;
	Souvenir de Bonn	Kümmler, 1922; Krenke, 1933; Bergann, 1962a
Abutilon pictum	Savitzii	Tilney-Bassett, 1963a
Acer negundo	Variegatum	Küster, 1919, 1927; Lakon, 1921; Kümmler, 1922; Funoaka, 1924; Thielke, 1948; Tilney-Bassett, 1963a; Manenti, 1975
Acer platanoides	Drummondii and Wittmachi	Küster, 1927, 1928; Pohlheim, 1983
Aichryson × *domesticum*		Tilney-Bassett, 1963a
Arabis albida	Variegata	Correns, 1919; Küster, 1919; Massey, 1928
Aubrieta decipens		Tilney-Bassett, 1963a
Aubrieta deltoidea	Aurea-variegata	Tilney-Bassett, 1963a
Bougainvillea glabra	Variegata	Tilney-Bassett, 1963a
Buddleia davidii		Tilney-Bassett, 1963a
Buxus sempervirens	Argenteo-marginata	Küster, 1919, 1927; Pohlheim, 1969b
Chelidonium majus	Variegata	Bergann and Bergann, 1960
Coprosma baueri	Variegata	Bateson, 1919, 1921; Küster, 1926, 1927; Sabnis, 1932; Tilney-Bassett, 1963a; Bergann and Bergann, 1983b
Cornus alba		Küster, 1919, 1927; Kümmler, 1922
Dianthus chinensis		Stewart, 1965
Diervilla florida	Variegata	Kümmler 1922; Küster, 1927
Elaeagnus pungens	Variegata	Küster 1927; Imai, 1935b; Bergann and Bergann, 1983b
Euonymus japonica	Albomarginatus	Bateson, 1919, 1921; Küster, 1919, 1926, 1927; Imai, 1935b; Dermen, 1960
Euonymus radicans	Variegatus	Küster 1919, 1927
Euphorbia pulcherrima	Eckes Rosa	Bergann and Bergann, 1960; Bergann, 1961, 1962c; Pohlheim, 1983
	Ruth Ecke and Bourbon Ecke	Dermen, 1960; Stewart, 1965
Ficus australis		Küster, 1927
Ficus parcelli		Küster, 1927
Ficus pumila	Variegata	Kümmler, 1922
Ficus radicans	Variegata	Bergann and Bergann, 1960; Tilney-Bassett, 1963a
Fraxinus excelsior	Argenteo-marginata	Küster, 1927, 1929
Fuchsia magellanica	Globosa and Gracilis	Küster, 1919; Kümmler, 1922; Tilney-Bassett, 1963a
Hedera helix	Marginata	Kümmler 1922; Küster, 1927
Helichrysum cordatum		Tilney-Bassett, 1963a
Helxine soleiroli		Tilney-Bassett, 1963a

Continued

Table 7.1 *(Continued)* Examples of dicots having the *GWG* sandwich structure.

Species	Cultivar	Investigators
Hoya carnòsa	Variegata	Bergann and Bergann, 1983b
Hydrangea macrophylla	Variegata	Sabnis 1923; Chittenden, 1926, 1927; Renner 1936a; Tilney-Bassett, 1963a
Ilex aquifolium		Focke 1877; Küster, 1919, 1926, 1927; Bergann, 1962a; Bergann and Bergann, 1983b
Kerria japonica	Aureo-variegata	Kümmler, 1922; Küster, 1927
Ligularia tussilaginea	Argentea	Küster, 1919, 1927; Funaoka, 1924
Ligustrum ovalifolium		Küster, 1919, 1927, 1932; Tilney-Bassett, 1963a
Ligustrum sinense		Stewart and Dermen, 1975
Ligustrum vulgare	Argenteo-variegatum Aureo-variegatum	Dermen, 1947d, 1960
Lonicera periclymenum	Quercina	Küster, 1927
Nepeta hederacea		Funaoka, 1924
Nerium oleander		Bergann and Bergann, 1983b
Nicotiana colossea		Honing, 1927
Nicotiana gigantea		Küster, 1919
Nicotiana tabacum		Dermen, 1960; Burk et al., 1964; Stewart and Burk, 1970
Pelargonium × *hortorum*	Golden Brilliantissima	Chittenden, 1926, 1927; Sabnis, 1932; Neilson-Jones, 1934; Imai, 1936b; Renner, 1936a
	Kleiner Liebling	Chittenden, 1927; Pohlheim, 1983
	Madame Salleron	Küster, 1919, 1927; Kümmler, 1922; Bateson, 1926; Chittenden, 1927; Renner, 1936a, 1936b; Bergann and Bergann, 1959
	Others	Küster, 1919, 1927, 1928; Kümmler, 1922; Imai, 1935b, 1936b; Tilney-Bassett, 1963a; Stewart et al., 1974
Pelargonium peltatum	L'Eléganté	Küster, 1919; Massey, 1928; Tilney-Bassett, 1963a
Peperomia glabella		Tilney-Bassett, 1963a; Bergann and Bergann, 1982, 1983a
Peperomia magnolifolia		Tilney-Bassett, 1963a
Peperomia obtusifolia		Bergann and Bergann, 1982, 1983a
Peperomia repens		Tilney-Bassett, 1963a
Peperomia scandens		Bergann and Bergann, 1982, 1983a
Plectranthus japonicus		Tilney-Bassett, 1963a
Prunus cerasifera		Küster 1932
Prunus persica		Demen, 1960
Rubus fruticosus		Tilney-Bassett, 1963a

Continued

Table 7.1 *(Continued)* Examples of dicots having the *GWG* sandwich structure.

Species	Cultivar	Investigators
Saintpaulia ionantha		Pohlheim, 1980b, 1983
Sambucus nigra	Albo-variegata	Küster, 1919, 1926, 1927; Renner, 1936a, 1936b; Thielke, 1948; Pohlheim, 1982b
Saxifraga stolonifera		Küster, 1919; Tilney-Bassett, 1963a
Sedum rubrotinctum		Pusey, 1962; Bergann and Bergann, 1984a
Solanum dulcamara	Variegatum	Küster, 1919
Solanum sisymbrifolium		Küster, 1919
Solanum tuberosum		Klopfer, 1965a; Simmonds, 1969
Symphoricarpos orbiculatus	Variegatus	Küster, 1927; Bergann and Bergann, 1960; Tilney-Bassett, 1963a
Ulmus procera	Argenteo-variegata	Küster, 1919, 1927
Viburnum tinus		Tilney-Bassett, 1963a
Vinca major	Variegata	Tilney-Bassett, 1963a

trichimera 'Weißrand Rosa' (*ABC*). Bud variations subsequently allowed Bergann to isolate new chimeras combining genotypes *A* and *B*, *B* and *C*, and *A* and *C*, as well as to isolate shoots wholly *A*, wholly *B* – which could not be propagated, and wholly *C*, and so increase the range of colours and cultivars. In respect of the ornamental bracts, four distinct colours were obtained from the two mutations and appropriate arrangement of layers as follows:

Pure white: *AAA* – 'Eckes Weiß'
White with pale pink centre: *AAB* and *AAC* – 'Trebstii alba'
Pink: *ABC* – 'Weißrand Rosa',
                                                     ABB and *ACC* – 'Eckes Rosa'
Red: *BBB*, *CCC* – 'Imperator', *BCC* and *BBC*

Some other trichimeras combine probable plastid mutation with inhibited growth. A likely early example mentioned by Darwin (1868) was a white-margined, dwarf *Pelargonium* 'Little Dandy'. Green bud variations were of two types, one with normal and one with dwarf growth (Bergann and Bergann, 1959). A better known case is the much studied sterile *Pelargonium* 'Madame Salleron', which has the following structure:

L I   Green, long internodes   (*A*)
L II   White, long internodes   (*B*)
L III   Green, short internodes   (*C*)

Bergann and Bergann (1959) accounted for the origin of 'Madame Salleron' in two steps. In 1830 a strongly growing variety *P.* × *hortorum* 'Fothergillii' (*AAA*) was grown in England. A mutation in L II gave rise to a vigorous white-margined sandwich chimera, which in 1855 was propagated as *P.* × *hortorum* 'Manglesii' or 'Mangles Variegated' (*ABA*). A second step took place in France in 1877 when a mutation to dwarf growth occurred in L III, owing to loss of internode elongation, coupled with the loss of ability to flower (*ABC*). Daker (1969) found $2n = 17$ chromosomes in 'Madame Salleron' instead of the expected $2n = 18$, leading him to suggest that loss of flowering might be related to chromosome loss; this would also account for the other growth changes that occurred at the same time. Bergann and Bergann were able to isolate the pure constitutuents *A, B* and *C*, together with new chimeras combining *A* and *B, B* and *C*, and *A* and *C* in various ways.

Inhibited growth of white tissue is quite common in *GWG* sandwich chimeras. When grown in the open, *Buxus sempervirens* 'Argenteo-marginata' had leaves with a clear white margin with just the occasional green edge, especially in the upper half. When grown in the greenhouse under humid, shaded conditions in a propagation box, the cuttings grew leaves with a broad green edge while the growth of the white L II was so inhibited that the green inner and outer layers came together (Pohlheim, 1969b). Another example of the influence of growth conditions was observed in *Acer platanoides* 'Drummondii'. Pohlheim (1983) compared the growth of leaves on long shoots, having an internode length of about 58 mm, with leaves on short shoots, having an internode length of about 6 mm. On the short shoots, the wide white margins covered about 34% of the leaf area compared with 11% on the leaves of long shoots. At the same time, owing to the inhibited growth of the white layer, flecks of green mesophyll developed on the edges of the white margin of leaves on long shoots, owing to the induced periclinal divisions in L I, were about nine times more frequent than on leaves of short shoots. By selecting two or more variants from clones of the poinsettia 'Eckes Rosa', the *Pelargonium* 'Kleiner Liebling' and one clone of *Saintpaulia ionantha*, each with a mutant L II, Pohlheim (1983) was able to show convincingly that the stronger the growth inhibition within the L II component, the more frequently and extensively did L I divide periclinally and so contribute to the development of marginal green mesophyll, as seen in the increasing frequency and size of flecks. Stewart and Dermen (1975) found the leaves of *Ligustrum sinense* had a narrow white border when grown at 16°C, and a wide border when grown at 26°C. When shifted from growth at 16°C to 26°C the new leaves responded by an increase in the growth of the white tissue from L II and a decrease in the growth of green tissue from L I and L III until they reached the proportions after growth continuously at 26°C. It appears as if increasing light intensity and raising temperature both favour the growth of mutant white tissue, but more studies are needed to fully clarify the causes of these environmental responses.

Stewart and Dermen (1979) have described a number of monocots as having the *GWG* sandwich structure (Table 7.2). The growth in length of the broad leaf buttresses, without the lateral expansion of the lamina observed in most dicotyledonous leaves, results in the typical linear leaf. In *Tradescantia albiflora* 'Albovittata' the single tunica formed the epidermis only, while periclinal divisions within the white cells of L II and the green cells of L III allowed the layers to push between each other and so become arranged into adjacent rows within the primordia, allowing the leaves to develop longitudinal stripes of varying width (Thielke, 1954; Clowes, 1957;

Table 7.2 Examples of monocots having the *GWG* sandwich structure.

Species	Cultivar	Investigators
Acorus gramineus	Variegatus	Imai, 1935b; Stewart and Dermen, 1979
Aglaonema modestum	Variegata	Stewart and Dermen, 1979
Aspidistra elatior	Variegata	Küster 1927; Stewart and Dermen, 1979
Ctenanthe oppenheimiana	Tricolor	Stewart and Dermen, 1979
Iris japonica	Variegata	Imai 1935b; Thielke, 1948
Iris pallida	Variegata	Sabnis 1932; Renner, 1936a
Ixolirion montanum		Stewart and Dermen, 1979
Liriope muscari	Variegata	Stewart and Dermen, 1979
Musa paradisiaca	Vittata	Stewart and Dermen, 1979
Pandanus veitchii		Küster 1927; Dermen, 1960; Stewart and Dermen, 1979
Sansevieria zeylanica	Craigii	Imai, 1935b
Stenotaphrum secundatum	Variegatum	Stewart and Dermen, 1979
Tradescantia albiflora	Albovittata	Thielke, 1954; Clowes, 1957; Neilson-Jones, 1969; Stewart and Dermen, 1979
Zebrina pendula	Quadricolor	Neilson-Jones, 1969

Neilson-Jones, 1969). Small buds developing in the broad axils of these leaves sometimes developed from within the region of a green or white row in the primordium and so grew as a wholly green or wholly white shoot respectively. *Zebrina pendula* 'Quadricolor' was similar but with broader green stripes. In a *GWG* shoot of *Aspidistra elatior* 'Variegata' L I contributed a green edge to the marginal mesophyll, but in addition the green L III in the core was often replaced by white L II, so that the green core was streaked with white stripes. Elsewhere in the marginal white zone, short green streaks revealed the places where L II had divided periclinally to dip into the mesophyll tissue fairly late in leaf ontogeny. The striping and streaking in these linear leaves is often so common that the underlying chimeral structure is easily obscured.

In some monocots, like *Musa paradisiaca* 'Vittata' and *Ctenanthe oppenheimiana* 'Tricolor', leaf growth was similar to that of the broad leaved dicots with the growth of a flat lamina from and at right angles to a central midrib. The main difference is that in these monocots the planes of cell division are extremely regular resulting in long parallel sectors from midrib to margin, whereas in most dicots the orientation of cell division is more random, resulting in sectors of various shapes. In both monocot species the contribution of L II and L III was highly variable and so the leaves had large sectors of green or white tissue running from midrib to margin without the regular marginal white skin and central green core so typical of dicots.

Stewart and Dermen (1979) assumed that *Dracaena deremensis* 'Warnecki' and *D. sanderana* are *GWG* chimeras, but Thielke (1948) had found that *D. deremensis* 'Bausei' ought to be interpreted as four-layered, *GWWG*, so perhaps the sandwich chimeras are really *GWGG*, Pohlheim

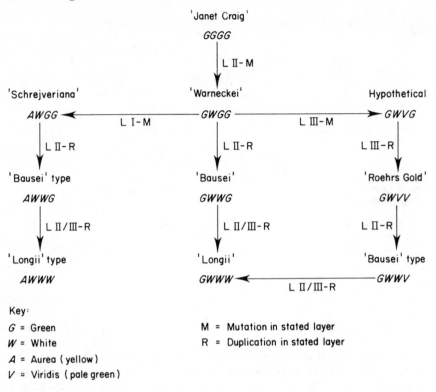

Key:

G = Green

W = White

A = Aurea (yellow)

V = Viridis (pale green)

M = Mutation in stated layer

R = Duplication in stated layer

Fig. 7.17 Proposed origin and constitution of the family of variegated-leaf sports of *Dracaena deremensis*. (Adapted from Pohlheim, 1982a.)

(1982a) supported this view after a detailed examination of a number of sports of *D. deremensis*. The complex interrelationship of the cultivars is summarized in Fig. 7.17. Essentially L I contributes to the leaf edges, and layers II to IV develop successive stripes of varying width from the leaf sides to the leaf centres. The chimeras, and hence cultivars, change by independent mutations in different layers and by duplications of layers forcing an existing layer downwards with respect to the growing-point, or inwards with respect to the leaf.

Eversporting conifers, *WG*

Several species of *Chamaecyparis* and *Juniperus*, belonging to the Cupressaceae family of Gymnosperms, possess variegated cultivars which were thought to be sectorial chimeras (Dermen, 1947d; Thielke, 1951, 1954). This was logical because the outer layer of the growing-point in many conifers does not have the character of a tunica (Johnson, 1951; Guttenberg, 1961). Nevertheless, Hejnowicz (1956, 1959) interpreted

Table 7.3 List of eversporting *WG* periclinal chimeras among conifers examined by Pohlheim.

Species	Cultivar	References
Chamaecyparis lawsoniana	Argenteo-variegata	Pohlheim, 1971a, 1971c
Chamaecyparis nootkatensis	Argenteo-variegata	Pohlheim, 1971a, 1971c
Chamaecyparis pisifera	Albo-variegata	Pohlheim, 1971a
Chamaecyparis thyoides	Variegata	Pohlheim, 1971a, 1971c, 1971e
Juniperus chinensis	Plumosa argenteo-variegata	Pohlheim, 1971a, 1971c, 1971e
Juniperus chinensis	Plumosa aureo-variegata	Pohlheim, 1971a, 71c, 71e, 80a
Juniperus chinensis	Variegata	Pohlheim, 1971a, 71c, 71e, 80a
Juniperus horizontalis	Variegata	Pohlheim, 1971a
Juniperus sabina	Variegata	Pohlheim, 1971a, 71c, 71e, 73a, 80a
Juniperus virginiana	'Triomphe d'Angers'	Pohlheim, 1971a

Juniperus sabina 'Variegata' as truly a periclinal chimera with a white L I and green L II, in which white sectors and shoots arose through periclinal divisions in L I. This interpretation was found to be valid for all eleven variegated cultivars of species examined by Pohlheim (1971a) (Table 7.3); there are also variegated conifers that are not chimeras. In every cultivar examined anatomically, Pohlheim (1971a, 1980a) found that occasional periclinal divisions occurred at the summit of the shoot apex causing duplication of L I and replacement of L II. When the cell descendants of the duplicated L I contributed to leaf development, or to the growth of an axillary bud, these tissues were wholly white and the previously green leaves became variegated with white sectors. The continuing growth of these mericlinal shoots either became increasingly white or grew towards green again. Once a shoot became fully white, it remained constant and no green sectors reappeared (Fig. 7.18). An analysis of the frequency of white sectoring showed that it was rare in the new shoot tips of axillary buds, but increased considerably with the distance from the place of origin (Pohlheim, 1971c). In *C. pisifera* 'Plumosa argentea' the white sectors arose more frequently in the lower than the upper side. The rarity of white sectors early in the development of a new shoot appeared to have two causes. On the one hand, periclinal divisions in L I were restricted to the summit of the shoot apex; they were not found either in the flanks or in the leaf or bud primordia. On the other hand, even at the apical summit, periclinal divisions appeared to occur very rarely in young shoots. It was these conditions that created the eversporting chimera. Periclinal divisions in L I at the summit of the apex regularly made the green shoots sport to white ones, but the white shoots eventually died out and so the growth of new shoots arose from the living green regions. Providing periclinal divisions had not already occurred in the main apical growing-point, and in young buds this was rare, the early growth of new buds was fully green as at this stage

Fig. 7.18 *Juniperus sabina 'Variegata'. An eversporting conifer with apparently green shoots, WG,* frequently sporting pure white shoots, *WW.* (From Tilney-Bassett (1963a). With the permission of the editors of *Heredity.*)

L I was very stable and all growth in volume depended on cell division in L II alone. Hence many buds arose in the axils of green shoots before any sporting had occurred, which happened mostly during their subsequent elongation.

Normally the variegated shoots rarely sported pure green, non-chimeral branches, but Pohlheim (1973a) found that after treatment of two year old cuttings of *J. sabina* 'Variegata' with X-irradiation at a dose of 0.25 to 0.75 krad, or, for better survival, a fractionated dose of 5 × 0.1 krad, L I was often displaced and pure green shoots regularly obtained. Evidently this treatment would be generally useful for isolating the core from other coniferous chimeras.

Comparison of the apical structure of these variegated conifers with the apices of some other species suggested that members of the Araucariaceen, Ephedraceen and Gnetaceen also had the potential for chimera formation, even if none had yet been reported (Pohlheim, 1971a).

8 Potato Chimeras

Potato shoots

Unlike cytochimeras or variegated-leaf chimeras, many sports have no revealing differences in size, shape or colour that mark them out as clear examples of chimeras. They appear like solid mutants, and it is only if they are observed to be unstable over time, or if they fail to propagate true to type, that one becomes aware of their chimeral nature.

It is probable that some chimeras are never recognized, but in commercial potato varieties a combination of several advantageous factors makes their identification unusually likely. The potato plant has the uncommon feature of producing two kinds of shoot. Above the soil surface are the green and leafy, vertical, aerial shoots. Below the soil surface are the colourless and bare, horizontal, stoloniferous shoots, terminating in the potato tubers. The tubers are not roots; they are large, swollen, storage shoots, on the sides of which are the eyes, containing rudimentary buds in the axils of leaf scars. Because they are shoots, potato tubers grow from apical meristems divisible into the same tunica and corpus layers as are found in the aerial shoots, and so any periclinal structure within the aerial shoot is conserved in the tuber too. The tubers faithfully propagate the whole plant and when they sprout, after over-wintering, the new aerial shoots that grow from the eyes faithfully reproduce the leaves and flowers of the previous vegetative generation.

Each potato variety is a clone in which every tuber is identical, and as each plant produces many tubers, a successful clone becomes widespread and numerous. So if the variety is a periclinal chimera, even if it is fairly stable, the large size of the clone inevitably results in some variation hinting at the underlying differences between the layers. Irrespective of the site and nature of the differential gene expression, the further analysis of a suspected chimera is greatly aided by the availability of the tubers. The tuber is an excellent experimental organ. The layer structure of buds developing from the germinating tuber is readily altered by eye-excision or irradiation treatment, following which the treated tuber may be allowed to grow up into a new plant in which the response to the treatment is observed. Moreover, as the tubers are clonal and numerous, the treatment can be practised on a sufficiently large scale to render meaningful results. The need to treat the less manageable leafy or flowering shoot directly is neatly avoided.

Methods of investigation

A rate of approximately 2×10^{-5} for the mutation from splashed pink to full pink in 'King Edward VII' was estimated by Howard (1959). Similarly, Heiken (1958) estimated rates from 1 to 8×10^{-5} for various kinds of leaf mutation, while bolters occurred with the higher frequency of about 2×10^{-3}. As potatoes are tetraploid, it is not surprising that many mutants are dominant, and therefore readily expressed, whereas recessive mutation must frequently remain hidden. Mutations have been induced by X- and γ-irradiation, but without leading to a significant development of new varieties (van Harten, 1978).

There are several ways of investigating the chimeral structures that arise from these mutations. The analysis of breeding behaviour is the best method for determining L II, but there are difficulties. As most potato varieties are pollen sterile, the character of L II cannot be determined by selfing. Instead, normal and mutant potatoes are each crossed with an appropriate fertile variety. When the germ layer is mutant the progeny from the two crosses are not identical, either they differ in phenotype or their segregation ratios differ; with tetraploid inheritance the simple characteristic Mendelian monohybrid or backcross ratios are not expected. A minor source of discrepancy is the estimated 2% of embryo-sac mother cells which Howard (1966a) traced back to L I in the 'Red King Edward' potato.

The method of eye-excision has been widely used for the determination of L III. As a rule a number of tubers belonging to the clone under test is cut longitudinally into two halves. One half is planted directly as the control, while the other half is planted after removing all the eyes, for which a special eye scoop may be used (van Harten, 1978). As further control, wholly altered and wholly unaltered tubers may be added. On those tubers with eyes removed, growth is dependent on the formation of entirely new, adventitious buds on the cut surfaces, and these usually develop from tissues derived from L III, which includes the inner cortex, vascular tissue and pith (Sussex, 1955). On many occasions adventitious buds have been reported to develop from the outer cortex of L II origin, but in Howard's (1970a, 1970b) opinion, this happened through faulty experimentation when buds were incompletely removed from the eyes. I doubt if this criticism is justified; there is no reason why the exposed surface after the buds have been removed should not sometimes be of L II origin, although the deeper the cut the less likely is this to be. A supplementary method of investigating L III is to develop adventitious buds on roots (Howard, 1964a; Miedema, 1967), but is less simple to operate than eye-excision.

In order to investigate the character of L I, and to isolate solid mutants from chimeras, van Harten (1972, 1978) used a range of experimental treatments to stimulate the production of adventitious buds from potato leaves, leaflets, and stem parts. Unfortunately, this species did not respond as hoped. The analysis of L I is therefore largely dependent on direct observation – as with coloured and russeted tubers, by interpreting layer changes – as when a thin-skinned chimera becomes a thick-skinned one, and by inference after confirming the genotypes of L II and L III.

The use of X-irradiation to alter or destroy the chimeral structure has been widely used (Howard, 1959, 1964b, 1967b; Heiken and Ewertson, 1962; Klopfer, 1965b; van Harten, 1978). The irradiation treatment usually consists of a range of X-ray doses focused on the rose end of sprouting tubers. The irradiation destroys many of the growing-point cells so that a new bud has to be reorganized from the surviving cells. Quite frequently there is no cell activity in one or two of the three layers so either the chimera structure is altered, or an entirely pure shoot arises. The kind and frequency of the changes tends to fall into a clear pattern. For example, Howard (1964b; 1967b) found about 50% of irradiated plants unchanged, about 45% of plants in which L I had been displaced by L II, and about 5% of plants in which L II had been replaced by L I. Other irradiated plants may have different, but equally characteristic, responses. Whatever the pattern, useful information can be inferred about the chimera structure.

As far as possible the investigation of chimeral structures should be made by several methods, and in the sections that follow particular attention has been drawn to conclusions based on experiments with adventitious buds – usually eye-excision, breeding, and irradiation treatment. In addition, as with all chimeras, a good description of the chimera and its spontaneous changes, and sometimes anatomical investigation, is invaluable.

Potato tubers

The classical work on the histology of the thorn apple, using cytochimeras, clearly showed that this species had a three-layered shoot apex. So it would be no surprise if the potato, within the same family Solanaceae, was similar, as was indeed found by the histological studies of Steinberg (1950), Sussex (1955) and Klopfer (1965a, 1965b). Moreover, cytochimeral studies demonstrated a three-layered structure by the existence of chimeras with a thin skin, a thick skin, and a sandwich structure (Table 6.1), and Howard et al. (1963) obtained some plants from the triploid hybrid Solanum × juzepczukii, which had a sandwich of a triploid epidermis, hexaploid pollen mother cells, and mostly triploid roots. The use of three layers in leaf development was evident in two GWG sandwich chimeras (Table 7.1), and the behaviour of many chimeral potato tubers is explicable on the same assumption.

Many potato sports affect the character of the tubers, especially their skin colour and texture, and so they involve the periderm. The analysis of these chimeras, together with histological studies, indicates that the periderm is developed from a cambium - the phellogen – which traces back to L I. The phellogen divides to give cells of the phellem towards the surface of the skin and to give the cells of the phelloderm towards the inside, the whole constituting the periderm (Esau, 1965). An exception to the development occurs in patches, particularly beneath the eyes and at the heel end where the origin of the phellogen traces back to L II (Howard, 1970a). It follows that if L I and L II differ in genotype with respect to a gene affecting the colour of the periderm, the patches are of a different colour to the rest of the skin, which aids the detection of chimeras.

The genetics of skin colour depends on at least six major genes controlling anthocyanin pigmentation (Howard, 1970a).

D is a basic gene which controls the development of a brownish-red colour in stems, inflorescences and sprout tips. Tetraploid potatoes homozygous for the recessive allele, *dddd*, have white tubers.

R develops a red colour in sprouts, and a rather weak colouring in stems. *R–D–* plants have tubers with a deep red phelloderm, but no pigment in the phellem and colourless eyes.

E develops a red colour in sprouts, stem and inflorescences. *E–D–* plants have tubers with a deep red colour in the phellem, and the colour may be deeper around the eyes. The eyes may still be coloured in the absence of *D*, when the rest of the tuber is white.

P is epistatic to both *R* and *E* and converts these colours to purple. It also gives purple sprouts and stems without *D, R* or *E*.

F is a gene for flower colour. *D–F–* plants have violet red flowers; *ddddF–* and *D–ffff* plants have white flowers; *P–D–F–* plants have purple flowers.

M restricts *E–D–* and *P–E–D–* tuber phellem pigmentation to areas round the eyes. It has no effect on *R–D–* and *P–R–D–* tuber phelloderm pigmentation.

It is apparent from this list that the precise colour and distribution of pigments in the tuber and elsewhere in the plant depends on specific combinations of genes and their interactions. The gene *M* has particularly interesting effects which give rise to splashed, hidden spotted, and spectacled colour patterns (Howard, 1970a). Not surprisingly, it is frequently difficult to distinguish from the descriptions of many potato varieties, associated with chimeras, which colour differences are related to the varying expression of a single genotype, and which to an underlying chimera structure. An additional difficulty arises from differentiation effects. When potatoes were propagated vegetatively after eye-excision or after treatment with X-rays, the next generation included some plants with white tubers from splashed parents, and spectacled tubers from hidden-spotted parents (Howard, 1962, 1966a, 1967a). In the case of 'King Edward VIII', Klopfer (1965b) assumed that the variety was a periclinal chimera with a white-splashed-pink skin and a white core. But according to Howard's explanation, which Klopfer did not discuss, the tubers have undergone only a temporaray change in gene expression from splashed-pink to white, which they maintain vegetatively, but which reverts again to splashed-pink from seed.

Ideally, the colour chimeras of potato tubers would be described in terms of their genotypes, and the actual genes which mutated to create the colour differences identified. Unfortunately, such detail is rarely known, and so I propose to describe the periclinal chimeras in the simpler terms of having the genotype for white or yellow (*W*), pink or red (*R*), purple or blue (*P*) skin colours. Even this classification is not without its problems. One problem, as already mentioned for 'King Edward VII', is that a red or purple skin is modified by gene *M* to give a splashed, hidden-spotted or spectacled phenotype, and this may become further modified to white. It is therefore not always clear whether the description 'white' refers to a true white or to a modified, coloured skin, and my use of white (*W*) probably includes examples of both types. An additional complication is that the tuber flesh that we eat, and derived from L III, is invariably white in appearance irrespective of its genotype. Its potential for producing a coloured periderm, and hence its chimeral classification, is determined by seeing the skin colour of the next vegetative generation of potatoes derived from the colourless tuber tissue by adventitious buds and, if these should still be white, the further confirmation of their remaining constant after propagation by seed. Such a careful analysis of white is usually lacking, and the description should therefore be regarded as tentative. With these reservations in mind, I have endeavoured to classify the potato skin colour chimeras (Table 8.1).

The majority of the chimeral tubers turn out to be the thin-skinned type in which a mutation has occurred in L I so that the genotype of L I differs from L II plus L III. In some chimeras the mutation has been from red to white, but I have found no description of a mutation from purple to white. In other chimeras the mutation is from white to red or white to purple. In 'Pink Arran Victory' the mutation was from purple to red and in 'Quarantaine Violette' from red to purple. In several instances the authors have recorded changes from a thin-skinned to a thick-skinned chimera arising through replacement of L II owing to duplication of L I – *WRR* to *WWR*. This change is sometimes reversed through displacement of L I owing to duplication of L III – *WWR* to *WRR*, as has been recorded for 'Red King Edward', 'Rote Holländische Erstling' and possibly 'Bovinia'. Both types of change may occur spontaneously, but the frequency is greatly enhanced by X-irradiation. In addition, reversion to one or other pure tissue, with loss of chimera structure, also occurs. The two varieties 'Blaue Eigenheimer' and 'Groene Rode Star' appeared to have become chimeras through a mutation in L II giving rise to the sandwich type of periclinal. These mesochimeras are probably less frequent than the thin-skinned chimeras owing to the lesser importance of L II than L I in the formation and colouring of the skin, and hence the greater difficulty in recognizing the mutation. An illustration of the effect of mutations and layer alterations on the colour of the skin and germination sprout is shown in Fig. 8.1. The strength of the evidence for the interpretation of the structure of these chimeras can be assessed from Table 8.1 by whether or not, in addition to the morphological description, the investigators have analysed L III by eye-excision, L II by breeding, and have also studied the effects of X-irradiation. In some cases, the structure of the chimera has been deduced by several criteria, in others by description alone, or by similarity with already proven chimeras.

Another kind of tuber chimera affects the texture of the skin. The 'Langworthy' potato has a thin, white, smooth skin whereas 'Golden Wonder' has a thick, brown, russet skin. When Crane (1936) propagated 'Golden Wonder' after eye-excision, some of the new tubers had a smooth skin just like 'Langworthy', indicating that 'Golden Wonder' had originated as a mutation from a smooth to a russet skin, comparable to the russet

Table 8.1 List of potato varieties that are periclinal chimeras for the skin colours yellow or white *(W)*, pink or red *(R)*, and blue or purple *(P)*.

Chimeral variety	Chimera structures	Experimental evidence			Investigators
		Eye excision	Breeding	Irradiation	
Bovinia	WRR WWR	+	−	−	Asseyeva, 1927
Bonte Rode Star	WRR	−	+	−	Dorst, 1952
Bonte Sport	WRR WWR	−	−	+	Howard, 1964b, 1970a
Burmannia	WRR	+	−	−	Van Harten, 1978
Désirée	WRR WWR	+	−	+	Van Harten, 1978
Noroton Beauty	WRR	+	+	−	Asseyeva, 1927
Wohltmann	WRR	+	−	−	Asseyeva, 1927
Merveille d'Amerique	RWW RRW	+	−	−	Asseyeva, 1927
Red Craigs Royal	RWW	−	−	−	Howard, 1959, 1971a
Red Gladstone	RWW	+	+	−	Howard, 1961b; Simmonds, 1965
Red King Edward	RWW RRW	+	+	+	Howard, 1958, 1959, 1962, 1966a, 1971a; Klopfer, 1965b, 1965c
Red Ulster Premier	RWW	−	−	−	Howard, 1959
Red Warba	RWW	+	−	−	Krantz and Tolaas, 1939
Rote Holländische Erstling	RWW RRW	+	+	+	Klopfer, 1965a, 1965b
Switez	RWW	+	+	−	Asseyeva, 1927
Unnamed purple	PWW	+	+	−	Asseyeva, 1927
Mindalny	PWW PPW	+	−	−	Asseyeva, 1927
Müllers Purple	PWW PPW	−	−	+	Howard, 1964b, 1970a
Tchugunka	PWW	+	+	−	Asseyeva, 1927
Pink Arran Victory	RPP	+	+	+	Howard, 1969a
Quarantaine Violette	PRR	+	−	−	Asseyeva, 1927
Blaue Eigenheimer	WPW	−	−	+	Klopfer, 1965a, 1965b
Groene Rode Star	RWR	−	+	−	Dorst, 1952

mutations that occur in apples. In a similar way the variety 'Great Scot' had sported the russet 'Sefton Wonder', and 'Up to Date' the russet 'Field Marshal'. Seedlings from these russet potatoes did not produce russet tubers, so the origin of the russet mutation could not have been in L II, and as eye-excision had shown that the mutation was not in L III, it followed that the varieties were periclinal for a russet L I and non-russet L II plus L III. Klopfer (1965b) has listed, and given the original references to, similar russet sports from several other varieties (Table 8.2). In each case the observed chimera had the mutation in L I because it is in this layer, which gives rise to the periderm, that russet is capable of being expressed.

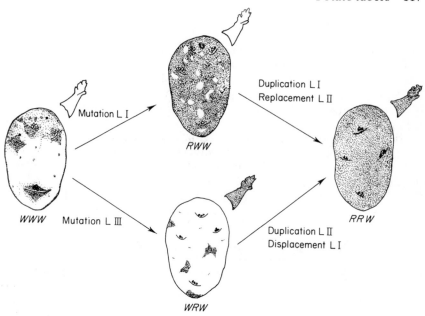

Fig. 8.1 Effect of mutations and layer alterations on the colour of the potato periderm and 'germination' sprout. *WWW* A white 'King Edward VII' potato with a strong splash of pink at the rose end and around the eyes. *RWW* Mutation in L I produces a red periderm, typical of 'Red King Edward', with white patches derived from the underlying white L II. *WRW* Mutation in L II does not alter the periderm which remains white, in this tuber without pink splashes, except where the underlying red L II has pushed through to give red patches, which are especially common at the heel end. *RRW* Duplication of the red layers of either L I or L II origin produces a red periderm without patches as both layers are now red. Confirmation of the genotype of L II is shown by the colour of the 'germination' sprout. (Adapted from Howard, 1959.)

Table 8.2 List of russet sports of potato that are periclinal chimeras for russet L I and normal L II plus L III. (Adapted from Klopfer, 1965b.)

Original variety	Chimeral variety	Investigators
Arran Comrade	Russet Arran Comrade	McIntosh, 1945
Bravo (smooth)	Bravo (russet)	Bolhuis, 1928
Burbank	Russet Burbank	Clark, 1930, 1933; Krantz, 1951; Rieman et al., 1951
Chippewa	Russet Chippewa	Rieman et al., 1951
Conference	Conference russet	Klopfer, 1965a
Great Scot	Sefton Wonder	McIntosh, 1945, Klopfer, 1965a
Langworthy	Golden Wonder	McIntosh, 1927, 1945; Crane, 1936; Klopfer, 1965a
Rural New Yorker	Russet Rural	Kotila, 1929; Clark, 1930; Krantz, 1951
Sebago	Russet Sebago	Weber et al., 1947; Rieman and Darling, 1947; Webster and Rieman 1949
Up to Date	Field Marshal	McIntosh, 1927, 1945; Crane, 1936; Klopfer, 1965a

Leaf mutants

The analysis of potato tubers, cytochimeras, graft chimeras, and variegated-leaf chimeras, makes it abundantly clear that potato shoots have a three-layered growing-point. Moreover, as is typical for three-layered plants, L I gives rise regularly to the leaf epidermis and sporadically to small areas of the marginal mesophyll. The margins and surface of the leaf blade below the epidermis is developed from L II, and the central region around and to the sides of the midrib from L III. Hence the most conspicuous leaf chimeras are likely to be those in which the mutation includes L II or L III.

Asseyeva (1927) described a mutant of 'Imperator', which she called the 'Kostroma' mutant, in which the leaves had an abnormal dissection and the flower corollas were also dissected. Her eye-excision technique showed that L III was unchanged, but it remained undecided whether L I, L II or both

Table 8.3 List of potato varieties that are periclinal for leaf mutants, bolters and wildings.

Chimeral variety	Probable structures	Experimental evidence			Investigators
		Eye excision	Breeding	Irradiation	
Arran Pilot					
Semi-bolter	NMN or MMN	+	−	−	Simmonds, 1965
Bintje Hasenlöffel	MMN	+	−	−	Klopfer, 1965b
Bintje Windenblatt	MNN	+	−	−	Klopfer, 1965b
Doon Star Ivy-leaf	NMM	+	+	+	Simmonds, 1965; Howard, 1966c; Howard, 1970a
Dunbar Rover					
Spinach-leaf	NMN or MMN	+	−	−	Simmonds, 1965
Eclipse					
Raspberry-leaf	NMN or MMN	+	−	−	Simmonds, 1965
Gladstone Wilding	NMN or MMN	+	−	−	Simmonds, 1965
Great Scot					
Dahlia-leaf	MNN or MMN	+	−	−	Simmonds, 1965; Howard, 1970a
Great Scot Wilding	NNM or NMM	+	−	−	Simmonds, 1965
Herzblatt	NMN or MMN	+	−	−	Klopfer, 1965b
King Edward					
Wilding	NNM or NMM	+	−	−	Simmonds, 1965
Kostroma	NMN of MMN	+	−	−	Asseyeva, 1927
Majestic					
Docken-leaf	NMM	+	+	+	Simmonds, 1965; Howard, 1965; Howard, 1970a
Majestic Feathery					
Wilding	NMM	+	+	+	Howard, 1969b
Majestic Holly-leaf	NMN or MMN	+	−	−	Simmonds, 1965; Howard, 1970a
Redskin Wilding	NMM	+	−	+	Howard, 1967c
Sharpe's Express					
Wilding	NNM or NMM	+	−	−	Simmonds, 1965
Sub-divided leaf	NMM	+	+	+	Heiken et al., 1963
Voronesh	MNN or MMN	+	−	−	Asseyeva, 1927

Key: N = Normal, M = Mutant layers.

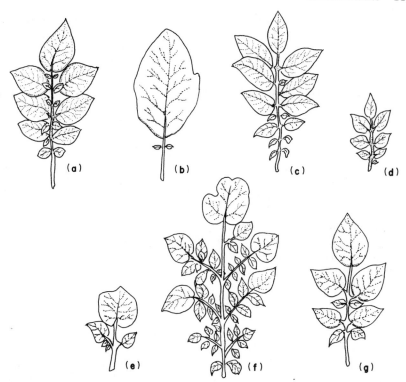

Fig. 8.2 Illustrations of a range of leaf mutants and wildings of the potato. **(a)** Normal leaf of 'Majestic'. **(b)** Majestic Docken-leaf. **(c)** Majestic Feathery wilding. **(d)** Majestic Holly-leaf. **(e)** Doon Star Ivy-leaf. **(f)** Subdivided leaf. **(g)** Redskin wilding. (Adapted from Howard, 1967b and 1970a.)

L I plus L II were mutant. Similarly, the 'Voronesh' mutant, derived from 'Wohltmann', had lost the bristle hairs from all organs, and in some flowers the corollas were white instead of purple, and were dissected. The effect upon hairs and flowers suggests that L I was probably mutant, but whether or not L II was mutant too remained undecided. Simmonds (1965) tested a number of leaf mutants by eye-excision and found examples both of L III normal and L III mutant, which defined the nature of this layer but not the full chimeral structure – it merely reduced the number of alternatives (Table 8.3). As the 'Great Scot Dahlia-leaf' mutant showed only slight modification through the contortion of young stems and the downward, marginal curl of the leaflets of mature plants, Howard (1970a) suggested that the mutation was confined to L I. Instead of pinnately divided leaflets, the 'Majestic Docken-leaf' mutant had more or less entire leaves, and the 'Doon Star Ivy-leaf' mutant had single, broad, lower leaves and compound palmate, very broad, irregular, upper leaves with broad leaflets (Fig. 8.2). These two, and the sub-divided leaf mutant, were tested by breeding, eye-excision and irradiation, proving that in all three L I was normal and L II plus L III mutant. Of course, as a mutation occurs in one cell of one layer it is probable that the original mutation occurred in L II and subsequently spread to L III creating the stable chimeras with L II plus L III mutant – *NMN* to *NMM*. The 'Majestic Holly-leaf' mutant was dwarf, with small, stiff, glossy, dark green leaflets, few tubers and no flowers; it seems probable that L II was affected. A radiation-induced dominant mutation for ivy-leaf in 'Burmania' proved not to be a chimera (van Harten *et al.*, 1973).

Klopfer (1965b) has described the sport-family 'Bintje'. The normal variety has the usual pinnately-divided leaves, whereas a pure mutant form has entire, spinach-shaped leaves, which he called 'Bintje Spinatblatt'. In between, a thin-skinned chimera with mutant L I had a deformed-looking, convolvulus-shaped leaf, whole at the tip and divided at the petiole end, which he called 'Bintje Windenblatt'. Finally, the thick-skinned chimera, with mutant L I plus L II, had entire leaves, broad and rounded at the tip and tapering towards the petiole, which he called 'Bintje Hasenlöffel'. After eye-excision, the two chimeras reverted to 'Bintje'. After X-irradiation, 'Bintje Hasenlöffel' reverted most frequently to the original variety – *MMN* to *NNN*, and a little less frequently, equally to 'Bintje Spinatblatt' – *MMN* to *MMM*, and to 'Bintje Windenblatt' – *MMN* to *MNN*. As a rule the changes were observed as mericlinal rather than completely altered shoots.

Bolters, wildings and feathery wildings

Three other kinds of shoot mutation, commonly found, are the bolters, wildings and feathery wildings. The bolters are taller, mature later, have longer stolons, and may flower more profusely than their parental varieties. The condition is thought to arise by a mutation to a short-day requirement for tubering from a variety tolerant to long days; they grow normally under short days (Howard 1970a). After eye-excision Simmonds (1965) obtained normal plants from the semi-bolter 'Arran Pilot', which indicated that it was not a solid mutant and probably had a mutant L II. Other bolters may have had both L II plus L III mutant, in which case eye-excision would not have distinguished possible chimeras from solid mutants.

The wildings (Table 8.3) are compact and bushy, with numerous, thin stems and many undersized tubers, there are fewer leaflets than normal and usually no flowers. Eye-excision gave rise to normal plants from a wilding 'Gladstone' and mutant plants from wildings of 'Great Scot', 'King Edward VII' and 'Sharpe's Express' (Simmonds, 1965). Howard (1967c) obtained mutant tubers after eye-excision from a 'Redskin' wilding, and a few normal plants after X-irradiation of the rose end of wilding tubers; he therefore concluded that L I was probably normal and L II plus L III mutant. A wilding 'Ackersegen', called 'Herzblatt', had heart-shaped leaves at the shoot tip while the lower leaves showed a gradation between entire leaves and those with little leaflets at the petiole end (Klopfer, 1965b).

The feathery wilding of 'Majestic', which is not a true wilding, was as vigorous as normal, had relatively thin stems, and distinctly narrower leaflets (Howard, 1969b). Eye-excision reproduced the feathery-wilding with even narrower leaflets suggesting that the mutation was not fully expressed in the chimera. Breeding also reproduced the mutation. Hence both L II and L III were mutant. Evidence that L I was normal came from the isolation of normal plants after X-irradiation.

Production of non-chimeral mutants

It is plain to see from the range of potato varieties examined that many mutations give rise to chimeral plants. This is not necessarily desirable. For practical plant breeding the production of non-chimeral mutants is important, especially when the traits for which mutations are desired cannot easily be selected for, as is the case with pest and disease resistance and yield. Many mutations are expressed in L I, but there is no really satisfactory method of easily obtaining a pure mutant plant from this layer. Mutations in L II cannot always be propagated by breeding, owing to sterility or self-incompatibility, and even if they can be successfully crossed the inevitable recombination of other genes may break down the unique qualities of the parental variety. Mutations in L III alone are difficult to spot and although they may be rescued by eye-excision this is not always successfully carried out. In all these instances, tests are needed to determine whether the new mutant is a chimera or not and, if so, which layer is affected. How much better if the formation of chimeras is avoided.

Methods of propagation, *in vivo*, to isolate the mutant tissue were not very successful, so techniques for propagation *in vitro* were developed using rachis, petiole, and leaflet disc explants (Roest and Bokelmann, 1976, 1980). Van Harten *et al.*(1981) irradiated explants of 'Désirée' prior to placing them on a culture medium. After sufficient growth of the uncontaminated explants, small cuttings were obtained from the developing adventitious sprouts, which were rooted and finally transplanted to soil. These plants were grown for three consecutive vegetative generations and were checked for induced mutations and for any sign of chimerism. There was an increase in mutation frequency with increasing doses of X-rays from 1.5 krad to 2.75 krad, with a spectrum at least as broad as for experiments *in vivo*. The average mutation frequency per tuber or per explant after propagation *in vitro* was increased to 74% compared with 24 to 38% after induction *in vivo*. Moreover, the frequency of chimeras *in vitro* averaged a mere 2%, about the same as the very best result from experiments *in vivo* using adventitious sprouts (van Harten *et al.*, 1972; van Harten and Bouter, 1973). As the method was relatively simple, not too laborious, and greatly reduced the frequency of chimera formation, it is very promising and should encourage the use of mutation breeding for the modification and improvement of potatoes.

9 Flower Chimeras

Early flower sports

New flower colours, and new colour markings on flowers, have always been of the greatest interest to gardeners, and especially to the breeders who are constantly on the look out for new sources of variation from which to extend and improve their range of cultivars. Throughout the nineteenth century the origin of many sports was enthusiastically recorded on the pages of the 'Gardeners Chronicle', or similar journals, and sports of many species were recorded by Darwin (1868). For example, the zonal pelargonium 'Rollisson's Unique' had purple flowers, which were known to sport to lilac, to rose-crimson, and to red. And on a red-flowered cultivar, red, rose, and crimson flowers were recorded all from the same truss, and on other trusses there were sectorial flowers half red, half lilac. Some of these sports would have been new mutations, but others were probably bud variations in periclinal chimeras of older origin.

Convincing evidence that flowers might quite often have a chimeral structure was published in a series of reports by Bateson on the outcome of propagation from root cuttings. The cultivar 'Hogarth', which had double flowers of a carnation scarlet colour, arose from a cross between a red-flowered *Bouvardia leiantha* and white-flowered *B. longiflora*. Another cultivar 'Bridesmaid' had double flowers in which the petals were pink on the outside and pinkish-white inside. After taking over sixty root cuttings of 'Bridesmaid', Bateson (1916) consistently obtained plants agreeing in every respect with 'Hogarth', from which he concluded that 'Bridesmaid' was a periclinal chimera having a pinkish-white skin over a core of 'Hogarth'. A similar finding occurred among several fancy pelargoniums. *Pelargonium* ×*domesticum* 'Escot' had white flowers with a large purplish red blotch on each of the petals, which tended to roll back. Root cuttings gave rise to plants with larger flowers, in which the peripheral areas were pinkish and the blotches were much redder on the petals, which did not roll back. Bateson (1921) intrepreted 'Escot' as a periclinal chimera in which the petals had a white skin of somewhat restricted growth over a pink core. 'Mrs Gordon' had white and pink flowers with lightly represented guide marks on the petals, whereas plants grown from the core had fully pink petals and deep crimson guide marks corresponding to 'Cardiff'. 'Pearl' was a white semi-double having small, purple patches in the area of the guide marks, whereas flowers derived from the core were heavily marked with red like 'Mme Thibaut' or 'Emmanual Lias'. Occasionally, the 'Pearl' flower had a red patch where the underlying tissue had broken through. Further examples of periclinal chimeras among fancy pelargoniums were said to be 'Duchess of Portland', 'Don Juan' and 'Chlorinda', and some zonal pelargoniums were also periclinal chimeras (Bateson, 1926). The zonal 'Salmon-fringed', also known as 'Skelly's Pride', had crumpled, glossy leaves, laciniated petals and sterile flowers, whereas root cuttings from this chimera had hermaphrodite flowers with normal petals and flat, matt leaves. Cassells and Minas (1983) have shown that the condition is not caused by a graft transmissable, infectious agent, but is a chimera, which tends to be unstable when propagated by meristem culture, and which reverts to the root cutting type when propagated by petiole explant culture. Similarly, 'Double New Life' had peculiar double flowers which were devoid of anthers, while its root cuttings reverted to producing normal hermaphrodite flowers. Finally, 'Kleiner Liebling' had minute, white-edged petals and no anthers, while its root cuttings gave rise to hermaphrodite flowers of normal size.

Root cuttings, or suckers from underground parts of the stem, which were assumed to be adventitious buds from the internodes, were used by Zimmerman and Hitchcock (1951) to demonstrate the periclinal nature of several rose sports. Thus one sport from the cultivar 'Briarcliff' had single pink flowers with crinkled petals and non-serrate leaves, whereas the adventitious buds reverted to the original cultivar with double pink flowers and serrate leaves. A second sport from 'Briarcliff' modified the skin into developing deep red flowers corresponding to 'Better Times'. Yet another chimeral sport was 'Souvenir', with yellow flowers, derived from 'Talisman', with red flowers, and still present as the core of 'Souvenir'.

For each of these cultivars the method of propagation by root cuttings had demonstrated the existence of a periclinal chimera in which the skin differed from the core. But Bateson had not distinguished between a thin or a thick skin, nor had he isolated the skin component as a solid, non-chimeral plant equivalent to his isolation of the core as a solid mutant, or solid normal plant. The next advance was made by Clausen and Goodspeed (1923) in their analysis of two chimeras in the tobacco. The chimeras arose as sports from pink to white, or carmine to pink, in the F_1 of crosses between the varieties *'macrophylla'* × *'cuba'* and *'purpurea'* × *'cuba'* respectively. Root cuttings established the periclinal nature of the sports, and the analysis of the F_2 progeny showed that in neither case had the germ line mutated. Hence the authors concluded that for each chimera, a single mutation from heterozygous to the homozygous recessive state of two separate genes affecting flower colour had occurred in the epidermal tissue, and that to a large extent the colour of the petal must be determined by pigment located in the epidermal layer. Similarly, *Myosotis arvensis* 'Star of Zürich' had blue flowers with a central white band on each petal, while 'Weirleigh Surprise' had the pattern reversed (Chittenden, 1927, 1928). The flowers bred according to the central band colour, which indicated that in these flowers the petals were derived from at least two layers – they developed their surface and marginal tissues from the epidermal layer, L I, and their central band from the sub-epidermal layer, L II. It was not long before it was realized that there might sometimes be a role for a third layer in flower formation, which led to an understanding of the full potential for periclinal chimeras in flowers and other organs.

Striped flowers

Pigmentation in the flowers of the Japanese morning glory is intense within the epidermal layer and lighter in the inner tissues, and in the stem it is restricted to the sub-epidermis. There is, therefore, the potential for discriminating between some types of chimera by observing the distribution of pigment in the stem and flowers. Full advantage of this possibility was taken by Imai (1934, 1935a), with the added bonus of triple flower colours. Sporadic mutation of the normal, blue flower gave rise either to a perfectly white flower, or to a flecked flower in which a white ground colour was covered by fine blue flecks. The flecking was actually caused by recurrent mutation from the recessive white to the dominant coloured allele, accompanied by a change in genotype from the homozygous recessive to the heterozygous condition. It is probably an example of a transposable element system (Nevers *et al.* 1986). On selfing flecked plants, approximately 3.5% of the progeny had blue, 4.4% mosaic and 92.1% flecked flowers. The mosaics were sectorial or periclinal chimeras (Fig. 9.1) which bore four types of shoot.

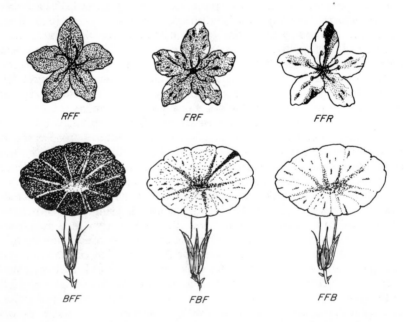

RFF   *FRF*   *FFR*

BFF   *FBF*   *FFB*

Fig. 9.1 Illustrations of periclinal chimeras for striped flowers in two species, each having a single full coloured layer in, from left to right, L I, L II and L III, and two flecked layers. **Upper row**: *Rhododendron obtusum* – *R* is a red and *F* a flecked layer. **Lower row**: *Pharbitis nil* – *B* is a blue and *F* a flecked layer. (Adapted from Imai, 1934, 1937.)

Type N   The stem was coloured and the flowers blue; the progeny segregated in a ratio of 3 blue:1 flecked. These shoots were either homogenous for the blue flower colour, *BBB*, or were thick-skinned chimeras with a hidden, flecked core, *BBF*.

Type I   The stem was green, and the blue flower colour was only slightly less intense than type N; the shoots bred as flecked. These shoots were presumed to have a thin, blue skin over a flecked core, *BFF*.

Type II   The stem was coloured, and the flowers were a light blue with intense blue flecks, and there was a white, discontinuous fringe around the petals; the shoots bred like type N. The periclinal structure was interpreted as having a sandwich of blue between two flecked layers, *FBF*.

The fringe around the petals showed that the colour underlying the flecked epidermis failed to extend completely to the margin, and the lightness of the colour showed by how much the blue was diluted by its position in the sub-epidermal layers.

Type III   The stem was green and the flecked flower had faintly blue coloured rays; the shoots bred as flecked. The periclinal structure was intrepreted as having a thick, flecked skin over a blue core, *FFB*.

The blue colour over the rays proved that L III did make a small contribution to the development of these flowers.

Like the flowers of the Japanese morning glory, coloured *Rhododendron* flowers are intensely pigmented in the surface layer and lightly pigmented within, and there are striped cultivars too. The striped flowers of *R. indicum*, which had red flecks on a white or tinged background, produced two common bud variations (Imai, 1935a). One kind of shoot had red flowers in which the guide marks on the upper petals were not evident. These shoots were assumed to be periclinal with a thin red skin over a flecked core, *RFF*. By contrast, the second kind of shoot had light red flowers with dark red flecks, fringed petals, and clear guide marks. These shoots were interpreted as having a sandwich of red between two flecked layers, *FRF*. The guide marks were prominent owing to the coexistence of red anthocyanin and green chlorophyll pigment within the same massive cells beneath the epidermis, but in the thin-skinned chimera, in which the red pigment was restricted to the epidermis, the guide marks were masked. Rarely, a third bud variation occurred with red flowers and dark red guide marks, and this shoot was presumed to have a thick red skin over a flecked core, *RRF*; it also occurred as a sport in fringed flowers by displacement of L I – *FRF* to *RRF*. The very attractive light red flowers, with dark red flecks, clear guide marks, and a continuous white fringe around the periphery of the corolla were highly esteemed and many cultivars such as 'Mme Morreau', 'Mme Joseph Vervane', 'Mme R. de Smet', 'Pres. O. de Kerkhove', 'Talisman' and 'Vervaeniana' were regularly cultivated and propagated by cuttings. Sports, like those of *R. indicum*, occurred in *R. ledifolium*, except that the fringing around the corolla was discontinuous and less conspicuous. Other flowering shrubs, in which Imai reported seeing chimeral cultivars with fringed flowers, were *Punica granatum*,

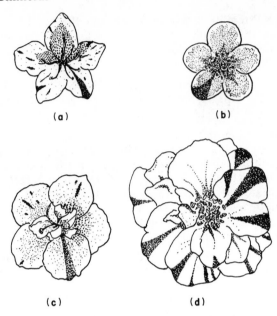

(a)   (b)

(c)   (d)

Fig. 9.2   Illustrations of periclinal chimeras for striped flowers. Each flower is a sandwich chimera with a full colour in L II between a white L I and L III. The flecks and stripes arise whenever the white initials in L I are displaced by the coloured initials in L II to develop into a lineage of fully coloured epidermis. **(a)** *Rhododendron ledifolium.* **(b)** *Chaenomeles lagenaria.* **(c)** *Rhododendron indicum.* **(d)** *Camellia japonica.* (Adapted from Imai, 1935a.)

Chaenomeles lagenaria and *Camellia japonica* (Fig. 9.2). The zonal pelargonium 'Mr Wren', which has white fringes to the red petals, remains true after meristem culture but dissociates into its red core after petiole explant culture (Cassells and Minas, 1983).

Another very interesting *Rhododendron* was the azalea, *R. obtusum* (Fig. 9.1). The cultivar 'Tokonatsu' was crossed with a white strain, in order to investigate flower variegation (Imai, 1935a; 1937), and three basic flower colours were obtained – red, flaked (another flecked form) and white. The red was further divisible into a darker and a lighter red, while the flaked was divisible into the standard flaked, which was white with red flakes, a lighter form with few red flakes, and a flaked red flower – red with darker red flakes. Formally, it would seem probable that there were multiple alleles for flower colour, which might well arise through the behaviour of a transposable element system (Nevers *et al.*, 1986). To Imai, what was important was that mutations could occur from flaked to red, or flaked to stable white, or from dark to light red, while the flaking itself arose from the recurrent mutation of an unstable white to red, and this too could alter in mutation frequency. It was therefore possible to obtain at least four series of periclinal chimeras – with red and flaked, with white and flaked, with red and white, and with red, white and flaked layers. As there were six possible chimeral structures for each pair of genotypes, six more for the combinations of all three colours together, and the three solid types, Imai (1935a, 1937) realized that the full potential amounted to 27 different combinations, and so was laid one of the cornerstones for our understanding of chimeral structures. Of course, not all of these chimeras were visibly distinct, especially as the inner core, derived from L III, was rather obscured, but several chimeras were recognizable including another example of a light red flower with red flakes and a white fringed margin, FRF, and similar chimeras were found in R. lateritium, which was usually grown in Japan as a bonsai. A further example of a variegated, chimeral flower was Chrysanthemum sinense 'Matsunami', which was described as having an epidermis of white and magenta stripes over a white core (Miyake and Imai, 1934).

A sport in Petunia hybrida arose through an epidermal mutation from deep to pale magenta to give a mericlinal chimera, in which up to one third of the circumference was periclinal for a pale skin over a deep magenta core (Bianchi and Walet-Foederer, 1974). In Petunia there is one gene with alternative alleles determining coloured (An) versus white flowers (an), which is epistatic to a second gene determining magenta (Rt) versus red (rt) flowers. A spontaneous mutation at the An locus in 'Roter Vogel' created a new mutable allele (an*), which was responsible for a variegated phenotype with the flowers having red spots on a white background. In some instances the variegated genotypes reverted to a full red, or to paler colours, or even to a stable white (Bino et al., 1984). When a reversion to a full colour occurred in the epidermal layer, the only layer in which anthocyanin was synthesized, the flower appeared red, but bred as if still variegated, showing that it was a periclinal chimera with a thin red skin over a genetically variegated, but phenotypically colourless, core, RVV. This property of forming epidermal, thin-skinned chimeras was utilized to determine the origin of plantlets from callus regenerating on leaf squares or leaf petioles. A heterozygous, magenta plant, An/an, Rt/rt, which sported a red branch, rt/rt, segregated into magenta and white progeny, showing that L II had not changed and that the red layer was limited to L I. When plantlets were regenerated from callus, they again developed red flowers like the original sporting branch, but these segregated into red and white flowered progeny after selfing. Hence the red flowering shoots developing from callus had developed solely from L I tissue losing their original chimeral structure. A similar result was obtained when branches with new flower colours – red, pale, white – were obtained on variegated plants. On selfing, the progeny segregated into variegated and white showing that L II was still variegated, but, with few exceptions, the regenerated plants developed flowers with the same colour as the branch from which they were descended and, if coloured, their progeny segregated into coloured and white showing that the regenerated plants again developed wholly from the epidermal layer.

A genetic analysis of the unstable genes for flower colour provided several examples of changes in the phenotype of the petals, with no change in the germ cells, owing to the mutational events being localized within a single layer (Doodeman and Bianchi, 1985). Other examples of the problems of mutation inducing the formation of periclinal chimeras in flowers occurred when an alteration in the germ layer, affecting the results of selfing or crossing experiments, was not accompanied by any change in the appearance of the petals. In both situations appearances were deceptive and the unwary breeder might easily be confused, unlike mutations inducing variegation in *Petunia* leaves in which the observed change was usually in L II, and where the alteration in the breeding pattern coincided with the alteration in the foliage.

Mutation breeding

For the most part, the practical aim of mutation breeding is to change a single character within an existing, commercially valuable stock, for instance flower colour, without altering the unique overall genotype. According to de Loose (1979), variation in *Rhododendron simsii* arose through the occurrence of an orange-flowered seedling in 1878, creating 'Apollo', and a carmine red-flowered seedling in 1948, creating 'Ambrosiana', while 'Knut Erwen' was a carmine red-flowered hybrid that arose in 1938. It was not until about 20 years later that the first spontaneous flower colour sport in azalea, 'de Waele's Favorite', which had a white edge to the carmine red flowers, came on the market. Similarly, 'Inga' arose as a white-edged sport six years after the origin in 1967 of the carmine red-flowered seedling 'Hellmut Vogel'. The rarity of spontaneous mutation explains why another aim of mutation breeding is to shorten the time of colour sporting, especially with economically important cultivars that have shown no sign of spontaneous sporting.

The spontaneous mutation in red 'Knut Erwen' gave rise to the white-edged, carmine red 'de Waele's Favorite' through loss of the ability to synthesize anthocyanin pigment in the epidermal layer. A further, irradiation-induced, mutation produced the sport 'Mevr. R. de Loose', in which the carmine red was changed to red through loss of the ability to synthesize flavonols in L I. Finally, the spontaneous white sport 'Mme de Waele' arose through loss of anthocyanin pigments from L II in addition to their earlier loss from L I. The white-edged colour sports, when crossed, behaved genetically in the same way as the original seedlings and so were periclinal chimeras with a thin white skin over a coloured core. De Loose (1979) was therefore able to use the chimera 'Mevr. R. de Loose', *WR*, to test the response to X-irradiation either after a single treatment of 2 to 3 krad, or after two or three recurrent treatments of 2 to 5 krad, in which shoots were alternatively irradiated, pinched out, regrown, and re-irradiated over two or three consecutive years prior to scoring the flowers. The general conclusion from these experiments was encouraging – by applying appropriate doses of irradiation, rearrangement of flower structure occurred up to fivefold more frequently after a single treatment than without irradiation, and from twenty- to fortyfold more frequently after recurrent treatment, although there was no obvious correlation with the actual dose. Of the changes in flower structure in 'Mevr. R. de Loose', about 93% were towards white, *WR* to *WW*, and 7% towards red, *WR* to *RR*. This suggests that L II suffered more severely from irradiation damage than L I, which contrasted with the effects of recurrent γ-irradiation upon 'de Waele's Favorite', in which flower colour changes in both directions were equally represented. A rewarding by-product of these experiments was the occurrence of a few new colour changes including a red with a very narrow white edge – 'Sierra Nevada', and a red flower with a very broad white edge. These changes presumably represent alterations in the amount of petal marginal tissue developed from L I rather than a change in the layer structure, as de Loose (1979) has not suggested that more than two layers are significant in petal development.

Readily sporting flowers

In contrast to *R. simsii*, in which spontaneous sports were all too rare, there are some species which sport frequently. One of these is *Chrysanthemum morifolium*, which is an outbreeding polyploid usually considered to be a hexaploid ($2n = 6x = 54$) with a variable somatic chromosome number (Langton, 1980). Many sports have arisen spontaneously by gene mutation, and others by chromosome loss. These have given rise to whole families of new cultivars differing in flower colour, size and vigour, all selected from original seedling cultivars, of which one of the most important commercially is the 'Indianapolis' (*IND*) family.

Although *Chrysanthemum* plants may be considered as having vegetative shoots based on a three-layered apex. Many yellow sports had chromoplasts either in the epidermis alone or in all internal tissues; there was no suggestion of more than two layers participating in petal development. Examination under the microscope showed that the petal colour was determined by two pigments. One pigment was a pink anthocyanin, which was found only in the vacuoles of cells of the upper epidermis. The cells in the lower epidermis and in the internal tissues of the petal retained the genetic potential for producing anthocyanin, but the gene responsible was not expressed. A second, yellow carotenoid pigment was located in the chromoplasts within the cytoplasm of all petal cells. The synthesis of both pigments could be independently stopped in either layer by specific mutation resulting in a possible 16 different genotypic combinations that might be expected to be expressed by up to eight different solid and chimeral phenotypes. Stewart and Dermen (1970b) investigated a sample of 16 *IND* cultivars to determine the nature of the variation between them. Their method was to establish the genetic potential for colour for each of the layers by following the known pedigrees of the cultivars, by examination of fresh petals under the microscope to locate the pigments, by encouraging the development of adventitious shoots after stripping off axillary buds to determine core genotypes, by observing the flower colours and sports therefrom, and by logical assumptions in the absence of contrary evidence.

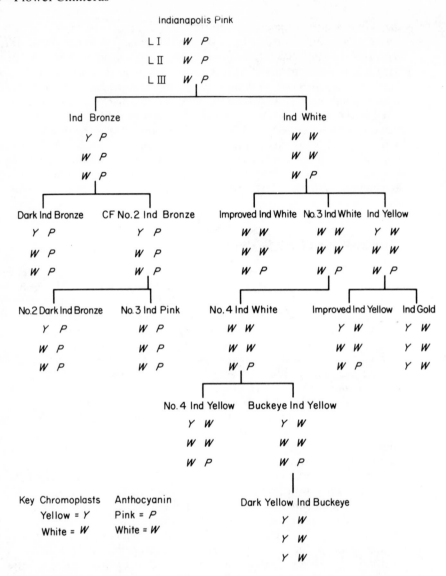

Fig. 9.3 Proposed origin and genotypes of the Indianapolis cultivars of *Chrysanthemum morifolium*. (Adapted from Stewart and Dermen, 1970b.)

The result was that half of the sixteen cultivars were thought to be chimeral for flower colour – L I differed from L II – and half of the remainder were thought to have chimeral shoots – L I plus L II differed from L III (Fig. 9.3). Four cultivars were trichimeras.

A few 'Indianapolis' chrysanthemums were later investigated by propagation from shoot tips, and by callus culture from petal segments or from the petal epidermis (Bush *et al.*, 1976). The cultivar 'CF No. 2 IND Bronze' was a chimera with carotenoids in the epidermis but not in the mesophyll. Yet in two thirds of plants originating from multiple shoots of a single apex, and in all plants originating from petal segment callus and, of course, in all plants derived from epidermal cultures, the petals had carotenoids in both epidermis and mesophyll. Evidently , there was a very strong tendency for the new shoots to be derived largely, if not entirely, from L I, which replaced L II. In addition to the frequent loss of chimeral structure, many plants developed abnormal, quilled, half quilled and incised florets, which rather suggests that the tissue culture process – involving the use of synthetic hormones and unnatural conditions – might have induced mutational changes or modified gene expression.

Examination of sections under the microscope, to determine the distribution of yellow chromoplasts, and progeny analysis to determine the character of L II, showed that a number of yellow-flowered chrysanthemums grown in Britain were chimeras too (Langton, 1980). The majority of yellow sports turned out to have a yellow L I and white L II layers, for example, 'Lemon Polaris', 'Yellow Arctic', 'Yellow Bonnie Jean', 'Yellow Hurricane', 'Yellow Marble' and 'Yellow Snowdon'. Apart from the colour, these plants were very like the cultivars from which they were derived. Occasionally a cultivar, such as 'Yellow Polaris', had carotenoid pigments in L II alone, while some cultivars like 'Golden Hurricane' and 'Golden Polaris' had carotenoid synthesis in tissues derived from both layers. These plants, with an altered L II layer, often had additional effects on such characters as flower size and flowering time owing, perhaps, to the multiple changes accompanying alterations in chromosome number or chromosome fragmentation that might have been responsible for the sporting.

A second species with numerous sports is the glasshouse carnation, *Dianthus caryophyllus*, to which the species *D. chinensis, D. fruticosus* and *D. plumarius* may have contributed (Farestveit, 1969). Although the carnation can be bred, the uniformity of each cultivar is maintained through readily propagated stem cuttings. Reports of very frequent colour changes after γ-irradiation of cuttings of 'William Sim' and several of its variants made it seem unlikely that the changes were due to mutation. Sagawa and Mehlquist (1957) therefore probed the matter further by treating shoots of various ages with doses of 2.5 krad or 5.0 krad X-irradiation. They irradiated the red-flowered 'William Sim', and two of its sports 'White Sim', and 'Pink Sim'. Only 3% of the irradiated branches of 'William Sim' bore flowers which differed from the normal red, whereas 82% of the shoots of 'Pink Sim' and 59% of 'White Sim' produced flowers which were totally or partially red instead of pink or white. Hence the two sports were much less stable than the parental 'William Sim', which was highly suggestive of a chimeral structure. Moreover, upon hybridization, the sports behaved exactly like 'William Sim'. As flower colour is developed in the epidermal layer alone, it became clear that the two sports were periclinal chimeras with a thin pink or a thin white skin over a red core – *PR* or *WR*.

A series of longitudinal sections through vegetative shoot apices of 'White

Sim' showed that irradiation often destroyed plugs of tissue. By four to eight days after treatment, a group of cells, including the epidermis and extending several cells deep, became necrotic. Over the next few weeks the dead cells were gradually sloughed off by enlargement and vacuolation of the cells just below the necrosis. The vacuolated cells then divided into a mass of initially disorganized meristematic cells, but by seven weeks after irradiation the new cells had differentiated into a reconstituted apex with a tunica and corpus. So, after the destruction of the the outer tunica, replacement came from the division of internal cells regenerating a new epidermis, and so accounting for the frequent flower change from a white or pink chimera to a solid red flower – *WR* to *RR*.

A second experiment with 2.5 to 40.0 krad γ-irradiation induced some new mutations, but again, in some cultivars, apparently 'dominant mutations' were actually histological changes of chimeras (Mehlquist and Sagawa, 1964). For example, the orange-flowered 'Harvest Moon' sported the lighter 'Cream Harvest Moon' and the deeper 'Tangerine', both of which proved to be periclinal chimeras. Similarly, 'Cardinal Sim', characterized by relatively smooth-edged petals, readily reverted to a typical 'William Sim' with serrated, fringed petals, and derivatives of 'Sidney Littlefield' proved to be as reversible as those from 'William Sim'.

Further evidence of the commonness of chimeras among carnation sports came from an extensive survey by Farestveit (1969). As flower colour is determined by various alleles of several genes (Mehlquist and Geissman, 1947), there is ample scope for many small deviations in colour which are discovered, propagated and marketed as new cultivars. Farestveit examined a selection of carnations marketed by Danish firms and divided them into three groups.

A   Sim carnations – all traced back to 'William Sim' either direct mutants, or mutants of mutants.

B   Non-Sim carnations – similar to the Sim type but of different origin.

C   Miniature carnations – these had smaller flowers, with several flowers on the same stalk, and were of more recent origin than groups A and B.

In order to stimulate changes, the shoots of rooted cuttings were irradiated with an electron beam from a linear accelerator to give an initial dose of 5 to 12 krad, they were then topped to encourage the growth of three to four lateral shoots and, when these were 3 to 5cm long, irradiated again with a dose of 5 to 6 krad. A few alterations, which were rare and dissimilar from other changes within the same cultivar, and which occurred in only 1.8% of treated plants, were new mutations. Other alterations, which occurred over a range of 31 to 85% of treated plants and were all of the same colour within a cultivar, were bud variations of chimeral shoots. More than half the cultivars tested proved to be chimeras and these occurred in all three groups (Table 9.1). Of the 24 Sim chimeras with two flower colours, 21 had a red flower colour as the core component, which was possibly the original 'William Sim', and 3 cultivars possessed a yellow core. In the Non-Sim and Miniatures too, the core was sometimes recognizable as one of the solid cultivars from which the chimera had presumably arisen.

Table 9.1 Frequencies of cultivars with solid or chimeral apices among three groups of carnations. A full list of the cultivar names and descriptions of their flower colours is published by Farestveit (1969).

Apical structure	Cultivar group		
	Sim	Non-Sim	Miniature
Solid	8	6	7
Dichimera	24	1	7
Trichimera	5	1	—

These observations showed that in six cultivars all three germ layers had a different genotype. For example, 'Sunset Sim' had an orange red flower which, after irradiation, frequently produced yellow flowers. When these yellow-flowered shoots were isolated and grown, red sectors occurred spontaneously as sectors of the flower, and when the irradiation dose was increased the orange red-flowered plants gave rise to shoots with red flowers directly. Once the red-flowered shoots were isolated no further change occurred. The progressive revealing of colours – orange red, yellow, red – clearly corresponded to the stepwise displacement of the damaged outer tunica by regeneration from the core – *OYR* to *YRR* to *RRR*. The occurrence of trichimeras means that although the flowers themselves are probably wholly derived from two layers, the vegetative shoots are three-layered, and may therefore retain in L III a genotype that is not normally represented in the flowers at all, yet may be brought into the flower at any time the apex is damaged and L I or L II has to be replaced.

Two other carnations with orange red-striped or orange flowers, which were sports of 'William Sim', were 'Jacqueline' and 'Jacky', these bred like 'William Sim' and so had a red L II. When cuttings were irradiated with 4 to 7 krad γ-rays, besides the normal flowers, there were branches with red flowers, and branches with salmon yellow, red-striped flowers (Péreau-Leroy, 1969). In order to follow what was happening, Péreau-Leroy (1974) used a 1% aqueous colchicine solution to induce a 244 and a 422 cytochimeral shoot apex in the 'Jacky' cultivar so that he could correlate changes in flower colour with changes in the ploidy of the apical layers. He found that the branches with orange flowers were unchanged cytochimeras, branches with yellow flowers always corresponded with L I ploidy, and branches with red flowers always corresponded with L II/L III ploidy (Fig.9.4). There was no evidence for three genotypes. In these two cultivars the orange flower was always a thin-skinned chimera with a single yellow layer over a genotypically red, but phenotypically colourless, core – *YRR*. The reason the skin was orange and not yellow in the chimera was attributed to a physiological interaction between the genetically distinct tissues in which 'yellow' L I was modified by 'red' L II, changing the expected, observed phenotype from yellow to orange, even though L II was not itself coloured, probably owing to an intercellular migration of an enzyme controlling pigment synthesis. The interaction must have been related to a specific condition as there was no comparable modification of the white

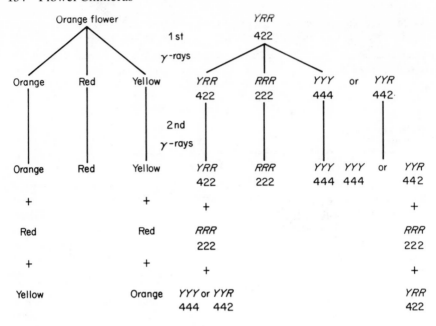

Key: Y = Yellow flower colour
      R = Red flower colour
      2 = Diploid layer
      4 = Tetraploid layer

Fig. 9.4 Changes in the shoots of the carnations 'Jacqueline' and 'Jacky' after irradiation treatment, as seen in the variation in flower colour, and as marked by the use of a 422 cytochimera. Left, the colours observed: right, the interpretation of the apical layer structure. (Adapted from Péreau-Leroy, 1969; 1974.)

petals to pink in 'White Sim', which also had a 'red' L II layer.

Although irradiation treatment is an effective way of developing a solid shoot apex from a chimeral one, and so overcoming the instability associated with chimerism, it has the disadvantage of itself inducing mutation. Dommergues and Gillot (1973) tackled this problem by inducing adventitious buds from apical or sub-apical *in vitro* cultures of 'White Sim', and by hormonal applications to the exposed surfaces of cut stems. Over 99% of the subsequent flowers were white and the remainder red. In order to test whether the whites were now solid white, or still chimeras, cuttings were irradiated with 7 krad γ-rays; the solid whites were stable, whereas the next vegetative generation of chimeral whites developed white and red flowers, and white with red sectors, owing to the numerous lesions induced in the apices. They found that after *in vivo* propagation about 75% were solid white and 25% chimeral, whereas after *in vitro* propagation of a small sample 100% were solid white indicating that regeneration had most often been from L I alone or L I plus L II together. Johnson (1980) also experimented with meristem cultures, and with cultures of physically macerated shoot tip explants, which he compared with irradiation by a dose of 6.8 krad γ-rays. Meristem culture of the chimera 'S. Arthur Sim' gave rise to 7% unchanged, chimeral flowers, 27% flowers derived from L I and 66% flowers derived from L II. Macerating the shoot tips of 'Braun's Yellow Sim', 'Dusty' and 'Pink Ice' gave a rather variable separation of chimeral, L I, or mixed flowers, but there were no L II shoots. Unfortunately, a low survival rate for the transference between sterile culture and greenhouse conditions made the interpretation rather vague. The chimera 'Dusty' proved particularly interesting. Flowers developed from L I alone were pale pink with red flecks, and from L II alone they were red, but the chimeral flowers were salmon pink with red flecks, together with a red sinus blotch where L II tissue had pushed through into an epidermal position. The cultivar was therefore another example of a chimera in which the petal colour developed from L I was modified by the different genotype of the underlying L II tissue.

Flower form

Not all flower sports affect pigments; a few, like the 'Salmon-fringed' pelargonium or the 'Briarcliff' rose, already mentioned, affect the form of flowers – petal shape, single or double flowers, fertile or sterile flowers – and sometimes the sports modify the leaves as well. An unusual, flower sport was described by Ikeno (1934) in *Erigeron annuus*. Normally the inflorescence is terminated by many daisy-like, composite flowers. Each little flower head possesses a number of inner, yellow, hermaphrodite and fertile disc-florets, and an outer circle of tongue-like, white, female and sterile ray-florets. Occasionally a mutation gave rise to an 'apetala' form, in which the white ray-florets were absent. The apetala form, which was recessive to normal, was unstable and frequently reverted to the normal both in somatic and generative tissues. Consequently, chimeras were frequently arising, giving the plants an extremely bizzare, ragged appearance. Sometimes the plants were sectorial for normal and wholly mutant flowering shoots, and many individual heads were sectorial with ray-florets arising just partly around the circumference, or in two or more short stretches, looking just as if chunks of florets had been plucked out. The gene controlling flower formation appeared also to be mutable during seed formation and so the progeny of apetalous plants were never constant.

An interesting flower sport that developed as a periclinal chimera was the 'micrantha' form of *Fragaria vesca* (Dahlgren, 1953, 1959, 1962; Pohlheim, 1978b). The petals were much smaller than normal, and were not in contact with each other and, as the sepals were short too, the yellow anthers and the carpels were visible in young flower buds. Crossing with normal strawberry plants failed to transmit the character, which suggested that the germ line was normal. The micrantha phenotype was not transmitted through grafts, which implied that it was not a virus infection, and so the most likely explanation was that the sport was a periclinal chimera with a dwarfing L I over a normal L II. Although vegetatively stable, rare reversion to normal supported the chimeral hypothesis.

The examples discussed leave one in no doubt that whenever flowers are improved by mutation breeding, or by the selection of spontaneous sports or, although not attractive, are maintained out of curiosity, many of the vegetatively propagated cultivars are likely to be chimeras (Broertjes and Ballego, 1967; Broertjes and van Harten, 1978), and unless appropriate steps are taken to select pure cultivars, the central problem of chimerism – periodic or frequent instability – will remain.

10 Fruit Chimeras

Mericlinal sports

There are a few plants, mostly trees and bushes, that are widely grown for the commercial value of their fruit. As the fruits are handled, transported, and sold directly to the consumer through retail outlets, an attractive shape, a smooth surface and a strong pleasing colour and odour, as well as a good taste, are essential ingredients for the successful variety. Not surprisingly, then, fruit sports are of more than passing interest, and have often been reported in the horticultural journals over the last two centuries. The history of the 'Bizzarria' orange in Italy, which produced a mixture of orange and citron fruit, goes back to the seventeenth century. Similarly, the 'Sweet-and-Sour' apple, in which some apples, and some bites of individual apples, taste sweet and others taste sour, has been discussed in the United States of America since the end of the eighteenth century (Gardner, 1944).

One important category of fruit chimeras is the extra large, giant-fruited sport, which is usually a cytochimera. Equally striking are the coloured sports, and sports with altered texture or ripening time, which are usually mericlinal chimeras; some of these might arise from chromosomal mutations, but mostly they are assumed to arise from spontaneous nuclear gene mutation. Several interesting examples of mericlinal fruit, in which the contrasting colours occupied large sectors were discussed in Chapter 3. There are many other cases in which the sectors are not quite so conspicuous as, for example, in the tomato (MacArthur, 1928) and squash (Hutchins and Youngner, 1952), as well as in the colours of orchard fruit (Drain, 1932); a few typical cases are illustrated in Fig. 10.1. Sectorials affecting foliage as well as fruit were described in the fig (Condit, 1928) and the olive (Roselli, 1972). Haskell (1965) analysed the differences in flavonoid pigments in an orange with orange and yellow sectors. Surprisingly, in comparison with the normal orange, the orange sectors revealed a lack of two of the eight flavonoid compounds, while the yellow sector revealed the further loss of a third flavonoid plus the addition of a new compound.

Frost (1926) suggested that many of the numerous bud variations on citrus fruit were due to the separation of components in long-existent periclinal chimeras, a possibility which is equally true for other fruit. Two types of tissue were observed in the same fruit of the 'Washington' navel orange (Shamel et al., 1926), and on two branches of a sour-sweet lemon, having orange-coloured flesh and a reduced acidity, typical sweet oranges were observed (Chapot, 1964). A greenish-yellow sector was reported on the dark red skin of the 'Kaiser Wilhelm' apple (Schmidt, 1942), and

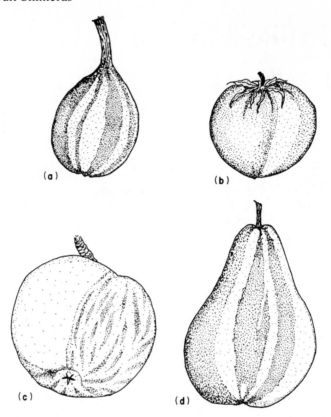

Fig. 10.1 Illustrations of mericlinal sports affecting skin colour in fruit. The lighter sectors arise either by new colour mutations, or by periclinal divisions of the initial cells of one layer displacing the adjacent initials of an alternately coloured layer in an existing periclinal chimera. As the sectors are rarely solid through all layers, they are mericlinal rather than sectorial. **(a)** fig; **(b)** tomato; **(c)** apple; **(d)** pear.

orange-yellow stripes on the red rind of 'Dobashibeni Unshu' a red colour mutant of the satsuma mandarin (Nishiura and Iwamasa, 1970). In all these plants sectoring may have arisen from new mutations, but might equally have arisen through the perforation of the skin by an underlying core of a periclinal chimera giving rise either to a sector or even a whole fruit or branch of the core type.

The 'Bartlett' pear is an interesting case (Gardner *et al.*, 1933). The developing fruit had alternating vertical stripes of yellow and green of varying width and mostly running from base to apex. Some trees appeared predominantly green and others predominantly yellow. A change from a broad to a narrow stripe frequently occurred at the nodes, and there were reversions to wholly green. The cultivar was considered to be a sectorial chimera, but the high incidence of regular striping and the general stability of the striping is unusual. An alternative possibility is the sorting-out of green and yellow plastids, but the leaves were not variegated. The reversion to green does not necessarily imply a mutable gene, so perhaps it was a periclinal chimera with yellow L I and green L II, in which L I regularly divided periclinally as well as anticlinally during the formation of the skin of the fruit, while retaining its customary stability during leaf development.

A bud sport in the tomato produced variant branches bearing nearly seedless fruit of inferior size and eating quality; the glandular hairs were smaller and the long trichomes fewer than normal (Lesley, 1929). It appears as if there was a mutation in the epidermal layer creating initially a mericlinal chimera that developed into a periclinal chimera in some branches. The 'Faupfull' peach likewise sported mutant short shoots that flowered later than normal and gave rise to fruit with a partial or complete loss of pubescence (Tsanev, 1976). Two early-flowering black currants – 'Daniel's September' and 'Laleham Beauty' – exhibited a characteristic leaf mosaic in all bushes and bore fruit of two distinct ripening periods. The early and late types were distinguished by their growth, flowers, fruit and seed content, berry weight and size, the length of the rachis and pedicel, and in the ripening time (Hughes, 1963). These contrasting features might have arisen from a mutation giving rise to a chimera structure, but attempts to fix the late-ripening form through shoot and root cuttings were unsuccessful and the chimeral hypothesis remained unproven. As is so often the case, a firm decision in favour of a sectorial or periclinal explanation, or between a chimeral and non-chimeral hypothesis, is thwarted by lack of sufficient information, but in other instances the evidence for a periclinal structure is more soundly based.

Periclinal sports

Grapefruit varieties are propagated by budding or grafting and so stable mutations are maintained, especially colour sports. Colour in the grapefruit, which may be expressed in the rind, septa and juice vesicles, is caused by lycopene and associated carotene pigments; unlike oranges, which have water soluble anthocyanins, the grapefruit pigments are not soluble in the juice. The cultivar 'Marsh', which is white with no pink colouration in the fruit, is the most common, nearly seedless, commercial grapefruit in the United States of America (Cameron et al., 1964). A limb sport on 'Marsh' gave rise to 'Thompson', which had pink flesh but no colour in the rind and, in its turn 'Thompson' sported 'Henninger's Ruby', which had both red flesh and red rind. A similar series occurred among seedy grapefruits. The white cultivar 'Walters' sported to 'Foster', which had a pink flesh and red rind, while 'Red Mexican' had a red flesh and red rind. Grapefruits often reproduce by seed through nucellar embryony – asexual reproduction of the mother's genotype – or by normal sexual reproduction. Citrus workers distinguish between the two types of origin by comparing the progeny with an extensive series of maternal tree and fruit characters. Nucellar seedlings of white 'Marsh' and red 'Henninger's Ruby' produced progeny with white or red fruit respectively. Similarly, the 'Red Mexican' and another cultivar 'Webb Redblush' produced seedlings that developed wholly red fruit and either had seeds or were nearly seedless like their respective parents. By comparison, the two sports 'Thompson' and 'Foster' behaved differently from each other. The nucellar seedlings derived from 'Thompson' developed wholly white fruit and those from 'Foster' wholly red fruit. Hence the colour of the progeny corresponded not to that of the maternal flesh, which was pink in both cultivars, but to that of the maternal rind. Apparently the rind was developed, like the seedlings, from L II, while the flesh was derived from another layer. In fact, histological studies showed that the juice vesicles arose as emergences from the inner surface of the carpel wall as result of repeated periclinal divisions in the epidermal cells, so that most of the flesh is of L I origin although some cells of L II origin are involved at their bases.

The genetic control of pigmentation in grapefruit was explained by Cameron *et al.* (1964) on the assumption that the development of red pigment is inhibited by a dominant allele, R, so that in heterozygotes, Rr, both rind and flesh are white. A mutation to the homozygous recessive state, rr, occurred in L I in 'Thompson' and L II in 'Foster' allowing red pigment to develop. Both cultivars appeared to have pink flesh, but this probably arose through some kind of interaction in which the white flesh was coloured by the red rind, or the red flesh was diluted by the white rind, perhaps through the movement of soluble precursors of the insoluble pigments. Hence 'Foster' appeared to have the periclinal structure L I – Rr, L II – rr, and 'Thompson' the reverse structure L I – rr, L II – Rr. From either of these periclinals the grapefruits with both red flesh and rind could have arisen by a duplication of the red layer. It also appears that in these fruit L III had the same genotype as L II, or its importance was slight.

Another limb sport of 'Thompson' gave rise to 'Burgundy' in which the flesh was red instead of pink. As the nucellar progeny again had white fruit, the authors (Olson *et al.*, 1966) assumed that L II was unaffected. Hence the change was in L I or L III. If in L I, there might have been a mutation to a new allele, r', producing a darker pigment than in 'Thompson'. If in L III, the change might have arisen by a rearrangement of layers such that a new L III was derived from L I. The latter mechanism could have created a sandwich of a white L II between red L I and L III, and the effect of two red layers might have reduced the diluting of the red colour to pink as seen in the flesh of 'Thompson' when only L I was red. A lemon with a reddish appearance and a decidedly pink colour to the flesh and juice (Shamel, 1932) probably had the same periclinal structure as 'Foster'.

The practically seedless 'Shamouti' orange has been known since the mid-nineteenth century (Spiegel-Roy, 1979). It probably originated as a limb sport on the common 'Beledi' orange, which has a rounder shape, narrower and smaller leaves, and more seeds. Nucellar seedlings from some 'Shamouti' trees were true to type, but from two trees all the seedlings were of the 'Beledi' form. Evidently, some trees were pure 'Shamouti' whereas others must have been periclinal with a 'Shamouti' epidermis and flesh derived from L I, and a 'Beledi' rind and nucellus derived from L II. Nucellar seedlings from 'Susuki Wase', a limb sport of satsuma, suggested that this cultivar was a periclinal chimera for an early maturing factor as the seedlings matured later than the parental clone. Histological development of the young fruit indicated that 'Susuki Wase' carried a common satsuma factor in the sporogenous and nucellar tissue, so presumably the mutation for earliness had occurred in L I or L III.

A sport in the normally wholly yellow 'Elberta' peach had the creamy-pink surface colour of the white-fleshed cultivars (Dermen, 1956). Internally, most of the fruits were bicoloured, with areas of yellow and white varying in extent and pattern, although some fruits were wholly white except for the yellow suture line. In other fruits the yellowing extended in some areas from skin to pit. Cytochimeral peaches showed that the suture was derived from L I, and that, when it contributed at all, varying amounts of tissue around the pit, were derived from L III, while the remaining tissue between the outer and inner epidermis was derived from L II. Clearly, the 'Elberta' sport was a periclinal chimera in which a mutation had occurred from yellow to white in L II, to give the yellow-white-yellow sandwich structure (Fig. 10.2). Another sport occurred in a white-fleshed seedling

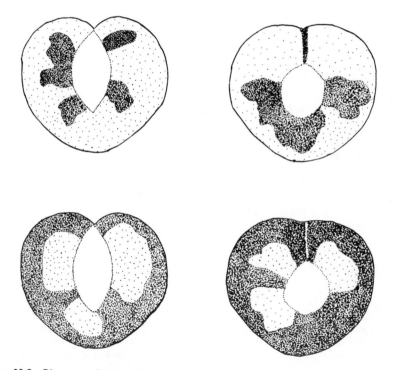

Fig. 10.2 Diagrams of chimeral peaches showing the irregular distribution of the layers contributing to the suture and to the bicoloured flesh. The upper pair has the sandwich structure L I yellow, L II white, L III yellow, and the lower pair has the reverse white, yellow, white sandwich structure. On the left of each pair the longitudinal section has the suture line at the back, and on the right the suture line is at the top. (Adapted from Dermen, 1956; and Yeager and Meader, 1956.)

from a cross between 'Eclipse' and a plant derived from the North Caucasus region (Yeager and Meader, 1956). The sporting branch was recognized by the appearance of some yellow fruit, which on cutting revealed irregular patterns of white and yellow flesh with the diagnostic suture white. Evidently this chimera, which was the reverse of the 'Elberta' sport, had a white-yellow-white sandwich structure. Yeager and Meader also referred to a sport in the reddish-black fleshed cherry 'Lambert' to a 'Rainbow Stripe' cherry, which was white except for a blood-red suture line. Presumably this chimera had the structure L I red, L II white and, if there was one, L III white.

Another peach sport, thought to be from the cultivar 'Shippers Late Red', had some fruit with the usual fuzzy skin, and others with fuzzy patches on an almost completely smooth skin. In general, these nectarine-like fruits were smaller, more highly coloured, and with a firmer flesh than the normal peaches. As he was not able to obtain a stable form of the sport, Dermen (1956) concluded that the mutation, which was assumed to inhibit fuzziness, was probably in the least stable L III layer, which contributes somewhat irregularly to fruit development. Moreover, he suggested, tentatively, that hair growth might be controlled by substances diffusing from L III into the epidermis. On the basis of these assumptions, the observed fruit were explained as completely fuzzy whenever L III failed to participate, as having fuzzy patches in segments where L III was lacking or too distant from the skin, and as having smooth patches or wholly smooth fruit whenever the mutant L III tissue was right beneath the skin, or had only a narrow strip of L II tissue between it and the epidermis. 'Shippers Late Red' was assumed, therefore, to be a thick-skinned periclinal chimera with a peach covering over a mutant, nectarine, core – PPN. By contrast, 'Angelo Marzocchella' appeared to have arisen by a mutation to nectarine in L I. This cultivar, widely grown in Italy, arose as a sport from the clingstone peach 'Vesuvio'. Selfing and crossing it reciprocally with the nectarines 'Panamint', 'Freedom' and 'Rose' showed that the germ layer was heterozygous for the recessive nectarine character (Collina et al., 1973); hence 'Angelo Marzocchella' was a thin-skinned periclinal chimera with a homozygous recessive nectarine L I, and a heterozygous peach L II and L III – NPP.

The commercially important 'Delicious' and 'Rome Beauty' apples produced frequent bud sports having an increased or decreased pigmentation, with or without a change in pattern from striped to solid colour, which indicated that some trees might be periclinal chimeras. This hypothesis was investigated experimentally by irradiating scions with γ-rays to a limit of 8 krad at the tip (Pratt et al., 1972b). As the light and dark blushed or striped sports of 'Rome Beauty' were resistant to change, and rarely lost their red pigments, it was suggested that they might have the generalized stable chimera structure NMM (normal, mutant, mutant), although it was not proven that they were chimeras. The 'Delicious' and its family of sports were more interesting. These differed from each other in the intensity and distribution of pigments within the epidermal and sub-epidermal cells (Table 10.1). Some of these sports were markedly unstable

Table 10.1 Distribution of anthocyanin in the skin from fruits of 'Delicious' apple and some of its presumed chimeral sports. (Adapted from Pratt *et al.*, 1972b).

Green Delicious:
Less than half of epidermal cells red. Few red cells in subepidermal tissues.

Delicious:
Few to most of epidermal cells red. Anthocyanin pigment 2 to 8 cells deep in skin, depending on striping or degree of blush.

Red colour sports (possible chimeras):
Few to most of epidermal cells red. Anthocyanin pigment 6 to 12 cells deep in skin, except in 'Norris'.

  Striped sports:
  Uneven distribution of colour – 'Red King'.

  Blushed sports: (Fig. 10.3)
  More uniform distribution of colour – 'Norris' – light blush. Most anthocyanin pigment 2 to 8 cells deep – 'Okanoma', 'Redspur', 'Richared', 'Royal Red', 'Starkrimson', and 'Vance'.

and gave clear evidence of an alternative colour hidden in another layer. The changes were interpreted by assuming that L I was normal, that the less stable sport had a mutant L II – *NMN*, or L III – *NNM*, and that the more stable were mutant in both L II and L III – *NMM* (Fig. 10.3). Another series of sports, this time derivatives of the apple 'Northern Spy', had increased anthocyanin pigments either in the epidermal layer alone, or in both epidermal and sub-epidermal layers (Table 10.2) (Pratt *et al.*, 1975). Breeding data, which suggested that red colour – either striped or blushed – was dominant over its absence and striped dominant over blushed, was used to determine the genotype of L II. The results supported the notion that these sports varied in the character of L II and might be periclinal chimeras – *MNN* or *MMN*, but L III was not tested and the explanations remain tentative.

The sweet-and-sour sport of 'Rhode Island Greening' bore mostly irregular fruits with some parts bulging out and tart in flavour, and other parts evenly rounded and sweet to taste. Some fruits were smaller than normal, perfectly shaped, and wholly sweet, and bud propagation from branches bearing such fruit resulted in trees bearing only small, rounded and entirely sweet apples. Gardner (1944) considered the apple to be a

Table 10.2 Distribution of anthocyanin-containing cells in the skin from mature fruits of 'Northern Spy' apple and some of its red, presumed thin- or thick-skinned, chimeral sports. (Adapted from Pratt *et al.*, 1975).

Predominantly striped fruit:
Red cells more or less discontinuous in outer layers of cells, depending on striping or degree of blush. Much anthocyanin pigment pale red.
  Anthocyanin mostly in outer 3 to 7 layers – 'Northern Spy'.
  Anthocyanin mostly in outer 4 layers – 'Bender Spy'.

Predominantly blushed fruit (possible chimeras).
Red cells composing most or all of outer 3 to 5 layers, less continuous 6 to 11 cells deep. Anthocyanin pigment bright red.
  Anthocyanin most in outer 4 layers – 'Kuppens Spy', 'Redwin Spy'.
  Anthocyanin in more than 4 layers – 'Crimson Spy', 'Farmer Spy'.
    and 'Schoharie Spy'.

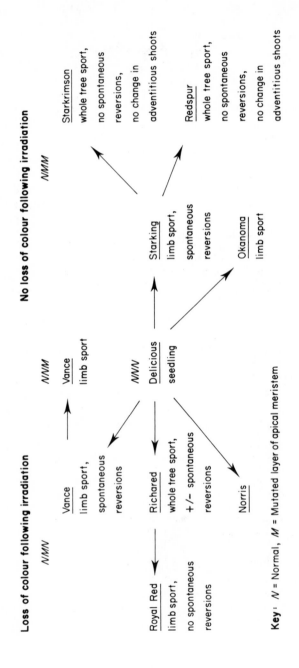

Loss of colour following irradiation

No loss of colour following irradiation

N/NN

N/NM

N/NM

N/NN

Starkrimson
whole tree sport,
no spontaneous
reversions,
no change in
adventitious shoots

Vance
limb sport,
spontaneous
reversions

Vance
limb sport

Starking
limb sport,
spontaneous
reversions

Redspur
whole tree sport,
no spontaneous
reversions,
no change in
adventitious shoots

Delicious
seedling

Richared
whole tree sport,
+/– spontaneous
reversions

Okanoma
limb sport

Royal Red
limb sport,
no spontaneous
reversions

Norris

Key: *N* = Normal, *M* = Mutated layer of apical meristem

Fig. 10.3 Diagram combining known origins of colour sports of 'Delicious' and hypothetical changes in chimeral structure, based on results or irradiation and literature reports. (Adapted from Pratt, *et al.*, 1972b.)

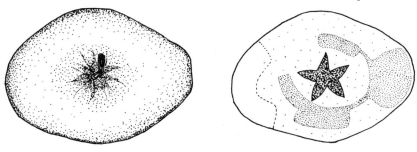

Fig. 10.4   Diagrams of the sweet and sour sport of the apple 'Rhode Island Greening'. On the left, the apple viewed from above is misshapen. On the right, a median transection shows the irregular distribution of the layers contributing to the flesh of the sandwich chimera. The surface is the outer epidermis. The edge of the star-shaped apple core is the inner epidermis. The unshaded zone to the left of the section is a portion of sour tissue derived from L I; the lightly shaded portion of sweet tissue is derived from L II; the heavily shaded portion of sour tissue is derived from L III. (Adapted from Dermen, 1948.)

Fig. 10.5   Illustrations of misshapen apples with irregular, longitudinal furrows and large lobes. They are assumed to be periclinal chimeras with a furrowed mutation in L III. (Adapted from Einset and Pratt, 1959.)

mericlinal or mosaic chimera, but Dermen (1948) interpreted it as a periclinal chimera. In Dermen's view, a mutation in L II had changed the growth from large and sour to small and sweet. Hence the sweet-and-sour apple was a chimera with a small, sweet tissue sandwiched between large, sour tissue. Much of the flesh was derived from L II and was sweet, whereas a sour taste in the outermost tissue was explained by assuming that part of the flesh was derived from L I, or that in places L II tissue was so thin that the sour L III approached the surface (Fig. 10.4). The limb sport giving rise to branches with small, sweet apples was assumed to arise through duplication of L II replacing L III – *NMN* to *NMM*.

Several sports of the apples 'Rhode Island Greening', 'McIntosh', 'Northern Spy' and 'Cortland' were reported as having mishappen fruit (Fig. 10.5) (Einset and Pratt, 1959). Their fruits were alike in having irregular, longitudinal furrows, but differed in the depth of furrows, the size of lobes, and the amount of russet and cracking. Unevenness in apples is often associated with lack of fertilization of some of the ovules leading to uneven seed development, but no such correlation was found in these cases. Russet was studied particularly in a russet-fruited sport of 'Stark' (Pratt, 1972). A few weeks after fertilization the epidermal cells of the fruit surface divided periclinally and died in wide areas. Russet then grew owing to the activity of a meristematic phellogen, arising in the most superficial cells, and

Table 10.3 Proposed chimeral structures of sports of the 'Rhode Island Greening' apple[1]. (Data from Pratt *et al.*, 1972a).

	S	R	P
Russet sports: Relatively unstable, hypothesized as *MNN* (*M* = Russet mutant)			
'# 2 *R* Palmer Greening'	S		
'Stark 287'			
'Whitney Russet King'			
Russet sports: Stable, hypothesized as *MMN*			
'# 1 *R* Palmer Greening'		R	
'# 3 *R* Palmer Greening'			
'# 1 *R* Datthyn Greening'			
Furrowed sports: Relatively unstable, hypothesized as *NNM* (*M* = Furrowed mutant)			
'Keukalaer Greening'	S	R	P
'# 2 *F* Datthyn Greening'	S		
'Cortland'			
'McIntosh'			
'Tucker McIntosh'			
Russet-furrowed sports: Relatively unstable, hypothesized as *NMM* (*M* = Russet-furrowed mutant)			
'Cornwall Greening'			
'McNicholas Greening'	S		
'Heinicke *A* and *B* Greening'	S	R	
'# 3 *RF* Datthyn'	S		

Examples of reversion to 'Rhode Island Greening', *NNN*: S = Spontaneous, R = Radiation stimulated, P = Pruning stimulated.

[1]New sports of 'Rhode Island Greening', indicated by #, originated in the orchards of Ralph G. Palmer or Eugene Datthyn; they are russet *(R)*, furrowed *(F)*, or russet-furrowed *(RF)*.

producing the corky periderm. Both spontaneously, and after treatment with up to 3 krad γ-rays, 4 to 8% of the sports reverted to give non-russet fruits like those of the parental 'Stark'. The rate of reversion was too high for new mutation, and so the sports were interpreted as periclinal chimeras with a mutant L I and normal L II and L III – that is they had a thin skin of russet-inducing tissue over a thick core of normal tissue – *MNN*.

After further analysis of the triploid 'Rhode Island Greening' apple, three kinds of mutation were reported with russet, furrowed, and russet-furrowed sports (Pratt, *et al.*, 1972b) (Table 10.3). Some russet sports were thought to have the periclinal structure *MNN*, and these could revert to normal whenever the mutant L I was displaced by L II – *MNN* to *NNN*. Other russets were highly stable, and in these russet L I had probably duplicated and displaced the normal L II – *MNN* to *MMN*. The furrowed sports were characterized by different degrees of lobing (Fig. 10.5) and, as they readily reverted to smooth and symmetrical fruit, they were assumed to have the relatively unstable structure *NNM*, in which reversion resulted from replacement of L III by duplication in L I or L II – *NNM* to *NNN*. The russet-furrowed sports were thought to arise from a separate mutation with its own phenotype, and not through a combination of russet and furrowed mutants. The russet-furrowed branches sometimes produced apples that had no russet in the furrows, but this was thought to occur when the mutant layer was shifted into the inner tissues and so no longer had any contact with the surface of the fruit. Hence the russet-furrowed sports were assumed to have the structure *NMM* and furrowed alone the structure *NNM*.

11 Other Chimeras

Mutant skins

Many examples of thin-skinned periclinal chimeras with a different or mutant epidermis have already been discussed. These have been classified in different chapters according to whether they arose by grafting, by polyploidy, or by gene expression causing variegation in the leaves, or changes in the colour or form of tubers, flowers or fruit. Inevitably, such a classification, for all its convenience, still leaves a few chimeras which do not fit, and so I feel it appropriate to place these in a rag-bag of left-overs. A rag-bag ought to be quite small, otherwise one might have good reason to think it had not been too well sorted. Hence, some examples that might have been included have been placed elsewhere because of their other associations. For example, *Antirrhinum majus* 'Wettsteinii', in which the leaves were affected, was discussed as a cytochimera because of its haploid skin and diploid core. Similarly, *Fragaria vesca* 'Micrantha' was discussed as a flower chimera, owing to the striking effect on the form of the flowers even though most other cases were concerned with flower pigments, and equally valid arguments apply to several other cases.

An important mutation of commercial interest is the thornless blackberry. Butterfield (1926) intrepreted a thornless sport of 'Cory ' as a periclinal chimera because adventitious shoots derived endogenously from the core were again thorny, whereas by the same argument, 'Austin', which remained thornless, was not a chimera. Darrow (1929) observed a thornless, chimeral sport of dewberry, and later Darrow (1931) obtained a thornless sport of *Rubus laciniatus* 'Evergreen'. He demonstrated that it was a thin-skinned periclinal chimera because its seedlings were thorny, showing that L II had not changed, and adventitious shoots from the roots were thorny, showing that L III had not changed either. These chimeras were not completely stable and were plagued by the vigorous bud reversions to thorny shoots in which L I, expressing the thornless genotype, was displaced by L II, expressing the thorny genotype. Stahl (1946) faced the same problem with the thornless 'Bowen' blackberry. The instability of these chimeras meant that they had to be constantly rogued, which limited their acceptabililty, and as they could not be bred from, they were of no value to the breeder. So, with these problems in mind, McPheeters and Skirvin (1983) set about obtaining non-chimeral thornless plants from the 'Thornless Evergreen' trailing blackberry. They propagated shoot tips of the chimera by tissue culture and obtained a proportion of *in vitro* regenerates with thornless stems, about half of which were dwarf and of solid thornless genotype, and half chimeral again. It appears that the pure thornless buds developed adventitiously from the epidermis, whereas there were no thorny shoots. The authors suggested that the medium served as a selective agent against adventitious thorny shoot formation, while favouring development of the thornless types. A portion of the thornless plants have flowered and set fruit, and so should now prove to be of direct use to breeders.

As its name suggests, *Laburnum anagyroides* 'Quercifolium' has four or five oak-like leaflets instead of the three smooth-edged leaflets of a normal laburnum leaf. A chimera structure became apparent when numerous progeny were all laburnum showing that L II had not changed (Bergann, 1954, 1955). The striking morphological change might indicate a mutation in L III, but the stability of the tree, which rarely reverted to laburnum, made it seem more likely L I contained the mutant layer (Pohlheim, 1978b).

The grapevine, *Vitis vinifera* 'Meunier', resembled 'Pinot noir' in essential characteristics, but was distinguishable by the dense white mat of hairs on the shoot tip and expanding leaves. Skene and Barlass (1983) used tissue culture methods to induce adventitious shoots from apical fragments. One type of regenerant resembled 'Meunier', another had tomentose leaves with comparatively hairless sectors, and a third was completely non-tomentose and resembled 'Pinot noir'. Evidently, 'Meunier' was a periclinal chimera. The formation of adventitious shoots on callus derived from fragmented apices had sufficiently disturbed the apices to allow buds to develop from a multicellular origin, either from residual periclinal regions, or from mericlinal regions, or solely from the tissues of the 'Pinot noir' core. The authors suggested that distortions in the leaves of the chimera were caused by the higher growth rate of hairless sectors.

In the poinsettias, the flowers are rather insignificant and it is for the ornamental bracts that the plants are generally grown. Stewart and Arisumi (1966) have described an interesting series of cultivars varying in bract colour owing to mutation from the heterozygous red, *Wh/wh*, to the homozygous recessive white, *wh/wh*, as follows:

'Paul Mikkelson'	red bracts	*RRR*
'Mikkelpink'	pink bracts	*WRR*
'Mikkeldawn'	white edge, pink centre bracts	*WWR*
'Mikkelwhite'	white bracts	*WWW*

'Mikkelpink' was recognized as a thin-skinned periclinal chimera because, when propagated, both by seeds and by adventitious shoots, the progeny failed to develop pink bracts. The cultivars 'Beltsvillepink' and 'Pink Ecke' were also thin-skinned periclinal chimeras with a white L I like 'Mikkelpink'. The bracts of pink chimeras showed occasional red streaks and sectors where the red tissue from an inner layer broke through the upper and lower epidermis. Preil and Engelhardt (1982) propagated the pink cultivar 'Annemie' by suspension culture, in which shoot tip callus was continuously shaken to separate out the cells, filtered, and replated on agar medium to re-grow. The new plants each developed either red or white bracts but never pink. This proved that the original plant was a chimera and that the cultural technique effectively separated individual cells and so dissociated the chimera into its two genotypes. In a similar manner, the cultivars 'Annette Hegg' and 'Brillant' were shown to be periclinal chimeras for internode length and leaf and bract shape.

Mutant cores

I have already discussed several examples of variegated-leaf chimeras in which severe distortions of the normal form were attributed to the inhibited growth of tissue derived from the sub-epidermal layer. Bergann and Bergann (1960) have described a curling sickness (Kräuselkrankheit) in a non-variegated poinsettia, in which the growth of the mutant L II was so inhibited that it could not develop normal palisade and spongy mesophyll tissue in the leaves. As a result, the centres of each leaf appeared to have a pale, rough band alongside the midrib, from the edges of which grew out large curly lobes of dark green healthy tissue derived from the normal L I. The plant was not attractive, and it seems likely that this kind of growth chimera is usually, promptly discarded.

A peculiar character that develops in several tomato cultivars is silvering (Grimbly, 1977; Grimbly and Thomas, 1977). This phenomenon results from a change in the sub-epidermis of the apical meristem and the cells derived from it. Towards the end of the normal division phase cell growth slows down, so that at maturity the cells are spread widely apart. The male and female floral parts develop normally until breakdown early in meiosis; the rare pollen is probably derived from small sectors of normal tissue. Although the plants cannot reproduce sexually, they can survive vegetatively. The evidence that silvering has the structure of a sandwich chimera is derived from adventitious buds. Those buds developing on cut stems all had green shoots with no trace of silvering, so L III was assumed to be normal, while those buds developing on leaf petioles, largely from the epidermis, were also fully green, so L I was assumed to be normal as well.

As a result of gamma irradiation, twelve mutant clones with compact growth were obtained in the 'Bramley's Seedling' apple, and approximately 500 trees of each clone were grown and checked for stability (Lacey, 1982). After treating the buds with a range of doses from 20 to 70Gy gamma radiation, seven of the clones proved to be stable as they kept their compact growth, whereas the other five clones were unstable and gave rise to a large proportion of apparently normal, taller trees. It therefore seemed probable that the stable clones were solid mutants and the unstable clones periclinal chimeras with normal L I and compact L II and L III. Confirmatory analysis of these tentative chimeras by other methods would be valuable.

12 Lessons and Prospects

Among vegetatively propagated plants, chimeras are widespread. The year by year large scale propagation of numerous individuals of a single cultivar raises the probability for spontaneous mutations to arise to near certainty, and whenever the rare mutations survive they are invariably preserved, and further propagated, as periclinal chimeras. Gene expression is all important and so, therefore, is the location of the mutation. The right mutation, in the right layer, soon becomes apparent from its obvious influence. In the wrong layer, the mutation remains hidden until revealed by a favourable bud variation, or by breeding or vegetative propagation. For some characters the right layer is L I, for others L II or L III. The prospects of controlling in which layer induced mutations occur is poor. On the brighter side, recent developments have shown that, like the induction of cytochimeras by colchicine, we can now use suitable mutagens, deliberately to enhance the frequency of plastid mutations leading to the prospect of extending the range of variegated-leaf chimeras. The induction of specific nuclear gene mutants remains a challenge, but at least their frequency can be enhanced by the prior selection for treatment of highly heterozygous individuals; and cell cultural techniques permit the treatment of numerous individuals, and the subsequent rapid propagation and marketing of any successes. When irradiation is used to induce mutation in higher plants, the readiness of bud variations to occur in the structure of known chimeras, suggests that these might be usefully incorporated into experimental procedures when seeking the most suitable dose.

Following mutation induction, there are still problems of recognition owing to the limitations of gene expression in chimeras, difficulties which are now well understood. For one character, such as plastid mutation, the ideal layer for gene expression in many plants, like L II in *GWG* sandwich chimeras, corresponds with the right layer for sexual transmission. For another character, such as a mutation in a gene controlling flower pigmentation, expression in L I and transmission in L II do not coincide. Consequently, mutations in L I may cause confusion when they are expressed but not transmitted, whereas hidden mutations in L II may cause confusion when they are not expressed but are transmitted. The understanding of chimeral structures and gene expression is an essential prerequisite for appreciating what is happening, and for adapting the correct strategy in handling these plants.

When chimeras are sought after, and often this is so because they display unique features that are not found in solid mutants, there are numerous ways of treating and manipulating them. The encouragement of bud variations, particularly by irradiation and propagation, to explore the variation in the layers, and so maximize the beneficial effects of gene expression, is routine. When chimeras are not sought, the same irradiation and propagation techniques may be used to dissociate chimeras into their constituent genotypes. Alternatively, the prospects of inducing mutations in cell and tissue culture avoiding the treatment of layered apices, and regenerating adventitious shoots from single or small groups of cells to produce solid mutants directly, is excellent. The necessary expertise is already well tested in some plants and there is every reason to expect that it will be successfully transferred to many more.

The use of suitably marked chimeras in the teaching of propagation is highly desirable. The lesson that propagation by seed, or by the induction of adventitious buds on root or leaf, may not always replicate the exact features of the parent, should be well learnt. Likewise, the finding that the multiplication of plants by the induction of callus on small explants of various organs does not reproduce the parental tissues with the precision of apical tip cultures, is most easily appreciated by the use of chimeras. The fact that the adventitious buds developing out of an irregularly shaped, crusty callus may each be derived from a specific layer, and even from a single cell, is not easily appreciated by mere visual inspection. Yet, unless this is the layer in which mutation has occurred, the main aim of propagation in some cases – to recover a solid mutant – may be totally thwarted. The finding that, in pelargoniums, chimeras may be distinguished from virus infections because the continuity of chimeras is maintained by apical tip culture but not by callus culture, whereas for viruses the result is reversed, may prove to have widespread validity.

The relative frequency of the different types of periclinal chimera that are maintained in cultivation is related to the localization of gene expression. A pleasing colour is highly valued, and whether we are looking at flowers or fruit, bracts or tubers, pigments are invariably concentrated in, and often exclusive to, the skin, and so we find the most common chimeras have an altered L I as compared with their parental cultivars. Often there is no gene expression in L II and L III, and the chimeral structure is not directly visible. Less often, when there is limited gene expression in tissues derived from L II and L III, an interaction between adjacent pigments of different genotypes creates a unique coloration absent from solid mutants. Alternatively, in petals, the contrast between a lightly coloured or colourless L I and a coloured L II creates a uniquely fringed flower. Another variation occurs in striped flowers when an inner layer, although itself colourless, regularly perforates the outer coloured tissue and is then expressed as flecks, sectors or stripes of contrasting colour. In a similar manner, a tuber skin of one colour may contrast with patches of another. Russet, thornless or hairless mutants are rather like the pigment mutants because they too are expressed in the skin and are therefore predominantly thin-skinned periclinal chimeras. Cytochimeras are often thin-skinned periclinal chimeras for the somewhat different reason that they are frequently selected by first looking for changes in cell and nuclear size, and chloroplast content, in the cells of the epidermal layer.

Variegated-leaf chimeras are less restricted than the anthocyanin pigment chimeras, because the expression of chlorophyll and associated pigments is normally spread right through the leaf and not localized in the skin. So a mutation in any one layer creates a visible colour contrast with the other two. In leaves, hidden chimeras arise through the nature of anatomical development and not through the limitations of gene expression. Hence, in some plants, the development of tissue from L I is restricted to a very stable epidermis, and chimeras with a differential epidermis are hardly recognizable. Thin-skinned chimeras are therefore most frequently found, and most interesting, when L I is not too stable, but periodically, or regularly, gives rise to conspicuous mesophyll tissue in teeth, lobes, or bands around the margin of leaves, or even to whole leaves. Sometimes this vigorous growth is strongly promoted by the highly inhibited growth of cells in the tissue immediately internal to it. The size and shape of leaves appears to be subject to a control over and above that of the individual layers from which it develops. Consequently, the mixture of two genotypes does not usually create a major developmental problem, and when the growth of one layer is severely restricted it is remarkably well compensated for by the supplementary growth of another. Not surprisingly, the greatest morphological adjustments occur in graft chimeras between different species or genera having very dissimilar leaves. In comparison with these, growth chimeras arising by mutation within a species, having a differential L II or L III but with no visible marker, are likely to prove quite difficult to detect and require careful investigation to confirm. Another kind of hidden chimera arises when a thick skin masks the differential core, and then the chimeras of ornamental value are those in which the masking is minimal, or non-existent owing to physiological or to structural factors such as the possession of a colourless hypodermal layer beneath the upper epidermis. Many variegated-leaf chimeras exist because they are not hidden chimeras. They are successful because, without any quirk of development, the colour contrasts occur in just the right layer for maximum effect. This is the case with the ubiquitous *GWG* sandwich chimera, in which the white green colour contrast corresponds with the normal development in many plants of the leaf margin from L II and the central tissues from L III. Conversely, the *WGW* sandwich chimera is rarely found, partly because it is unlikely to arise by direct mutation in two separated layers, or to arise as a bud variation from any other kind of chimera, and partly because the white tissue occurs in the two layers L I and L III where it is most likely to be hidden. Finally, what is so conspicuous about the graft chimeras is the complete absence of any sandwich structure; all are thin or thick-skinned chimeras. Obviously, this is related to their mode of origin in which callus from the first species grows over callus from the second; the callus of one species appears unable to burrow into the middle of another to make a callus sandwich. So we see that the chimeras we find in cultivation are not random, but are closely related to their mode of origin, to gene expression, and to variation in anatomical development.

Bud variations in chimeras vary greatly in frequency. On the one hand, there are many variegated-leaf chimeras – perennials, shrubs and trees –

which, because of the visibility of the white-over-green chimeral structures, can be seen to be extremely stable over long periods of time. They may be cultivated and propagated with little fear of any breakdown. On the other hand, there are variegated-leaf chimeras which quite commonly throw off green and white shoots, and these require constant pruning to maintain the correct condition. In addition, there are the eversporting conifers, and flowering plants with a similar behaviour, which have a mode of development which combines the short term instability of individual shoots with the long term stability of the chimeral structure of the plant as a whole. These are bud variations that reflect the degree of stability in the layers of the growing-point, but we may also witness degrees of stability, or instability, in the growth of specific organs – leaves, flowers and fruit. Stability in organ development is not necessarily correlated with stability in the growing-point, and there are many examples of chimeras in which the contribution of the two or three layers is highly variable as between the green and white zones of individual variegated leaves, or between the contrasting colours of striped flowers or fruit, yet the chimeral structure of the whole plant is very constant. When we analyse the variation, the most common cause is found in periclinal divisions within the cells of a tunica layer causing replacement of the cells of the adjacent inner layer or corpus tissue. The displacement of the cells of an outer layer by cells within is a far less frequent change, especially in the growing-point. When displacement does occur in the growing-point, this may often be attributable to surface damage requiring the regeneration of an epidermal layer from the cells below. Within the developing organ the growing cell population is not so constructed as in the tightly layered growing-point, and so pockets of high meristematic activity may encourage cell derivatives of one layer to push a wedge through and so displace those of another. Cell damage may be induced by treating apices to varying doses of X- or γ-irradiation. The death of cells in the outer tunica leads to their displacement by the cells below. Equally, cell death within an inner tunica or corpus, leads to their replacement by the cells above. It is suggested that irradiation increases the frequency of bud variation rather than altering the type, but the pattern of recovery from irradiation damage is variable between chimeras, and as yet we do not enjoy the prospect of precisely controlling the treatment in order to maximize regeneration from specific layers at will. It remains a most useful technique for changing the growing-point from one chimeral structure to another, but where solid shoots are required, it does not have the same control as adventitious bud techniques, and carries the danger of inducing unwanted new mutations. The rather different use of irradiation for the induction of single mutations on plants with minimal disturbance to a unique, existing genotype – mutation breeding – is likely to continue in importance.

Comparisons between chimeras, whether graft chimeras, cytochimeras or variegated-leaf chimeras, have repeatedly shown that the essential development of the plant is often not greatly altered by the differences in genotype between the layers. The chimera grows, branches, flowers and fruits much like its non-chimeral parent. Hence, in most cases, the contrasting phenotypes, corresponding to the contrasting genotypes, can be used as clear signals for the pathways of ontogeny. Chimeras have shown very consistently that the concept of histogens is invalid; there is no direct relationship between the way a cell differentiates and its place of origin in the apical meristem. What matters is where it finds itself in the growing and differentiating organ. Chimeras have shown that there is great flexibility between plants in the choice of route for the development of comparable organs. In some plants, there is no stable tunica at all, yet in others a clear separation exists between two, three or even four layers. Irrespective of the apical organization, leaves, flowers, and fruit are still formed. Even within an individual plant, the demand upon the apical layers may vary with perhaps three tunicas and a corpus contributing to the stem, three tunicas to the leaf, and two tunicas to the flower. Between species, and to a lesser extent between organs within a species, there is great variation in the demand upon individual layers. In many species, L I is steadfastly anticlinal in division covering all the plant surfaces with an epidermis and no more, yet in others regular periclinal, as well as anticlinal, division results in extensive areas of sub-epidermal tissue of L I origin – most noticeable in the margins of some leaves and the sutures of some fruit. Similarly, tissues of L II origin vary from little more than a thin covering beneath the epidermis to practically the entire growth of an organ. Except where a fourth layer is occasionally involved, L III supplies any further demand for tissue not fulfilled by L II, and is therefore complementary to L II. Within the individual plant, chimeras show that great flexibility is tolerated without sacrificing the ultimate form and function of the organ. A discussion of the finer details of anatomical development is beyond the scope of the present book, but for the interested reader there is much further information in the many articles to which I have referred. As pointed out particularly by Stewart and Dermen (1975, 1979) and Bergann and Bergann (1984c), there are many fascinating features of plant development that are only revealed in the behaviour of chimeras and the observations from chimeras are not contradictory, so it is surprising that many anatomists have been reluctant to make use of them. I concur with these authors in regarding chimeras as indispensable materials in the study of plant development, and it is my hope that this short book will increase awareness of their merits. Of course the importance of looking at sections in plant anatomy is paramount, but such studies, not least of all in teaching, could be greatly enhanced by the visual impact of suitably chosen chimeras, and the living plants are incomparably more useful for revealing and quantifying the less frequent events.

Chimeras have also played a role in physiological experiments, and some examples of these have been referred to. As with anatomical studies, there appears to be a lack of awareness of the potential of chimeras, and of the wide range of sources available. Variegated-leaf chimeras have long enjoyed a role in schools for the elementary demonstration that both light and chlorophyll pigments are essential for the synthesis of starch by photosynthesis. They deserve to be used more widely and more often. Mutation breeding plays an essential role in the improvement of vegetatively propagated plants, and in the process of inducing mutations chimeras frequently arise from which the solid mutant has then to be isolated. This has come to be known as the chimera problem. Of course, in these circumstances where solid mutants are specifically required, chimeras are a problem. Nevertheless, in all other respects, chimeras are just the opposite; they are plants of considerable use in research and teaching, of great fascination, and highly appreciated for their ornamental qualities. The case for chimeras is proven – *Nemine contradicente* – their prospects are excellent.

References

Ackerman, W. L. and Dermen, H. (1972). A fertile colchiploid from a sterile interspecific *Camellia* hybrid. *J. Hered.*, **63**, 54–9.

Arisumi, T. (1964). Colchicine-induced tetraploid and cytochimeral daylilies. *J. Hered.*, **55**, 254–61.

Arisumi, T. (1972). Stabilities of colchicine-induced tetraploid and cytochimeral daylilies. *J. Hered.*, **63**, 15–18.

Asseyeva, T. (1927). Bud variations in the potato and their chimerical nature. *J. Genet.*, **19**, 1–26.

Asseyeva, T. (1931). Bud mutations in the potato. *Trudy prikl. Bot. Genet. Selek.*, **27**, 135–217. (Russian)

Bain, H. F. and Dermen, H. (1944). Sectorial polyploidy and phyllotaxy in the cranberry (*Vaccinium macrocarpon* Ait). *Amer. J. Bot.*, **31**, 581–7.

Baker, R. E. (1943). Induced polyploid, periclinal chimeras in *Solanum tuberosum*. *Amer. J. Bot.*, **30**, 187–95.

Barrett, H. C. (1974). Colchicine-induced polyploidy in *Citrus*. *Bot. Gaz.*, **135**, 29–41.

Barrett, H. C. and Hutchinson, D. J. (1982). Occurrence of a spontaneous octoploid in apomictic seedlings of a tetraploid *Citrus* hybrid. *Proc. Int. Soc. Citriculture*, Tokyo, Japan 1981, Vol. 1, 29–30.

Bartels, F. (1956). Zur Entwicklung der Keimpflanze von *Epilobium hirsutum*. I. Die im Proembryo ablaufenden Zellteilungen bis zum beginnenden 'herzförmigen Embryo'. *Flora*, **144**, 105–20.

Bartels, F. (1960a). Zur Entwicklung der Keimpflanze von *Epilobium hirsutum*. III. Wachstumstendenzen, die zur Entwicklung eines Laubblattprimordium und Keimblattes führen. *Flora*, **149**, 225–42.

Bartels, F. (1960b). Zur Entwicklung der Keimpflanze von *Epilobium hirsutum*. IV. Der Nachweis eines Scheitelzellen wachstums. *Flora*, **150**, 552–71.

Bartels, F. (1966). Der Sproßscheitel von *Epilobium hirsutum*. *Ber. dtsch. bot. Ges.*, **79**, 309–12.

Bateson, W. (1916). Root-cuttings, chimaeras and 'sports'. *J. Genet.*, **6**, 75–80.

Bateson, W. (1919). Studies in variegation. I. *J. Genet.*, **8**, 93–9.

Bateson, W. (1921). Root-cuttings and chimeras. II. *J. Genet.*, **11**, 91–7.

Bateson, W. (1924). Note on the nature of plant chimaeras. *Studia Mendeliana*, 9–12.

Bateson, W. (1926). Segregation. *J. Genet.*, **16**, 201–36.

Bauermeister, W. (1969). Untersuchungen über die Peridermbildung bei einer neuen Laburnocytisus-Chimäre, ihren Stampflanzen und Rückschlagstypen. *Wiss. Z. Päd. Hochsch. Potsdam*, **13**, 139–55.

Baur, E. (1909a) Das Wesen und die Erblichkeitsverhältnisse der 'Varietates albomarginatae hort' von *Pelargonium zonale*. *Z. Verbungsl.*, **1**, 330–51.

Baur, E. (1909b). Pfropfbastarde, Periklinalchimären und Hyperchimären. *Ber. dtsch. bot. Ges.*, **27**, 603–5.

Baur, E. (1910). Pfropfbastarde. *Biol. Zbl.*, **30**, 497–514.

Baur, E. (1911). Entgegnungen. *Z. Bot.*, **3**, 198–202.

Baur, E. (1930). *Einführung in die experimentelle Vererbungslehre.* 7./11. Aufl. Gebrüder Borntraeger, Berlin.

Behera, B., Tripathy, A. and Patnaik, S. N. (1974). Histological analysis of colchicine-induced deformities and cytochimeras in *Amaranthus caudatas* and *A. dubius*. *J. Hered.*, **65.**, 179–84.

Beijerinck, M. W. (1900). Over het entstaan van knoppen en knopvariations bei *C. Adami. Verh. Akad. van Wetensch. Amsterdam*, **9**, 336.

Beijerinck, M. W. (1901). Über die Enstehung von Knospen und Knospen-varianten bei *C. Adami. Bot. Z.*, **59**, Pt 2. 113–18.

Beijerinck. M. W. (1908). Beobachtungen über die Enstehung von *C. purpureus* aus *C. Adami. Ber. dtsch. bot. Ges.*, **26a**, 137–47.

Bergann, F. (1951). Haberlandts Crataegomespilus-Studien ein Beitrag zur Frage der vegetativen Hybridisation. *Züchter*, **21**, 245–53.

Bergann, F. (1952). Über das Auftreten einer bisher unbekannten Laburnum–Form an der Pfropfchimäre *Laburnum adamii. Flora*, **139**, 295–9.

Bergann, F. (1954). Praktische Konsequenzen der Chimärenforschung für die Pflanzenzüchtung. *Wiss. Z. d. Karl-Marx-Univ. Leipzig*, **4**, 281–91.

Bergann, F. (1955). Einige Konsequenzen der Chimärenforschung für die Pflanzenzüchtung. *Z. Pflanzenzüchtung*, **34**, 113–24.

Bergann, F. (1956). Untersuchungen an den Blüten und Früchten der Crataegomespili und ihre Eltern. *Flora*, **143**, 219–68.

Bergann, F. (1961). Eine weitere Trichimäre bei *Euphorbia pulcherrima* Willd. *Biol. Zbl.*, **80**, 403–12.

Bergann, F. (1962a). Über die Beteiligung des 'dermatogens' an der Mesophyllbildung. Paradigmatische Schichtenverlagerungen an den Blättern albovariegater Periklinalchimären. *Wiss. Z. Päd. Hochsch. Potsdam*, **7**, 75–86.

Bergann, F. (1962b). Über histogenetisch bedingte Anomalien bei der generativen Aufspaltung periklinal Chimären. *Wiss. Z. Päd. Hochsch. Potsdam*, **7**, 87–94.

Bergann, F. (1962c). Über den Nachweiss zwischenzelliger Genwirkungen (Partnerinduktionen) bei der Pigmentbildung in den Brakteen der Periklinalchimäre *Euphorbia pulcherrima* Willd. 'Eckes Rosa', *Biol. Zbl.*, **81**, 469–503.

Bergann, F. (1967a). The relative instability of chimerical clones – the basis for further breeding. *Abh. Dt. Akad. Wiss. Berlin*, **2**, 287–300.

Bergann, F. (1967b). Mutations-Chimären: Rohmaterial züchterischer Weiterbehandlung. *Umschau in Wiss. u. Techn.*, **24**, 791–7.

Bergann, F. and Bergann, L. (1959). Über experimentell ausgelöste vegetative Spaltungen und Umlagerungen an chimärischen Klone, zugleich als Beispiele erfolgreicher Staudenauslese. 1. *Pelargonium zonale* Ait. 'Madame Salleron'. *Züchter*, **29**, 361–74.

Bergann, F. and Bergann, L. (1960). Über die sogenannte Kräuselkrankheit der Poinsettia und die Beteiligung des 'Dermatogens' bei der Mesophyllbildung. *Flora*, **149**, 331–44.

Bergann, F. and Bergann, L. (1962). Über Umschichtungen (Translokationen) an den Sproßscheiteln periklinaler Chimären. *Züchter*, **32**, 110–19.

Bergann, F. and Bergann, L. (1982). Zur Entwicklungsgeschichte des Angiospermblattes. 1. Über Periklinalchimären bei *Peperomia* und ihre experimentelle Entmischung und Umlagerung. *Biol. Zbl.*, **101**, 485–502.

Bergann, F. and Bergann, L. (1983a). Zur Entwicklungsgeschichte des Angiospermenblattes. 2. Über die Blattmusterbildung bei meso- und diektochimärischen Formen von *Peperomia*-Arten, insbesondere über die Beteiligung des 'Dermatogens' an der Mesophyllbildung. *Biol. Zbl.*, **102**, 403–29.

Bergann, F. and Bergann, L. (1983b). Zur Entwicklungsgeschichte des Angiospermenblattes. 3. Über unmaskierter Binnenfelder in den Blattspreiten periklinalchimärischer Buntheit von *Elaeagnus pungens, Coprosma baueri, Ilex aquifolium, Hoya carnosa* und *Nerium oleander. Biol. Zbl.,* **102,** 657–73.

Bergann, F. and Bergann, L. (1984a). Zur Entwicklungsgeschichte des Angiospermenblattes. 4. Über Periklinalchimären bei *Sedum rubrotinctum* R. T. Clausen. *Biol. Zbl.,* **103,** 147–71.

Bergann, F. and Bergann, L. (1984b). Gelungene experimentelle Synthese zweier neuer Pfropfchimären – die Rotdornmispeln von Potsdam: + *Crataegomespilus potsdamiensis* cv. 'Diekto', cv. 'Monekto'. *Biol. Zbl.,* **103,** 283–93.

Bergann, F. and Bergann, L. (1984c). Zur Entwicklungsgeschichte des Angiospermenblattes. 5. Über die Anlegung von Blättern und Blütenorganen im Lichte klassischer und moderner Histogeneseforschung. *Biol. Zbl.,* **103,** 655–75.

Bianchi, F. and Walet-Foederer, H. G. (1974). An investigation into the anatomy of the shoot apex of *Petunia hybrida* in connection with the results of transformation experiments. *Acta Bot. Neerl.,* **23,** 1–6.

Bino, R. J., Bianchi, F. and Wijsman, H. J. W. (1984). Periclinal chimerism in *Petunia* demonstrated by regeneration of plants from the mutated epidermal layer. *Heredity,* **52.,** 437–41.

Blakeslee, A. F. (1941). Effect of induced polyploidy in plants. *Amer. Nat.,* **75,** 117–35.

Blakeslee, A. F. and Avery, A. G. (1937). Methods of inducing doubling of chromosomes in plants. *J. Hered.,* **28,** 393–411.

Blakeslee, A. F., Bergner, A. D., Satina, S. and Sinnott, E. W. (1939). Induction of periclinal chimeras in *Datura stramonium* by colchicine treatment. *Science,* **89,** 402.

Blakeslee, A. F., Satina, S. and Avery, A. G. (1940). Utilization of induced periclinal chimeras in determining the constitution of organs and their origin from the three germ layers in *Datura. Science,* **91,** 423.

Blaser, H. W. and Einset, J. (1948). Leaf development in six periclinal chromosomal chimeras of apple varieties. *Amer. J. Bot.,* **35,** 473–82.

Blaser, H. W. and Einset, J. (1950). Flower structure in periclinal chimeras of apple. *Amer. J. Bot.,* **37,** 297–304.

Boeke, J. H. and van Vliet, G. J. C. M. (1979). Postgenital fusion in the gynoecium of the periclinal chimera *Laburnocytisus adamii* (Poit.) Schneid. (Papilionaceae). *Acta Bot. Neerl.,* **28,** 159–67.

Bolhuis, G. (1928). Geschubde and ongeschubde Bravo's. *Landbouwkd. T.,* **40,** 760–6.

Bond, T. E. T. (1936). Disease relationships in grafted plants and chimeras. *Biol. Rev. Camb. Phil. Soc.,* **11,** 269–85.

Bornmüller, J. (1932). Über Rückschlagbildungen an *Crataegomespilus. Mitt. dtsch. dendrol. Ges.,* **44,** 75–80.

Brabec, F. (1949). Zytologische Untersuchungen an den Burdonen, *Solanum nigrum-Lycopersicum. Planta,* **37,** 757–95.

Brabec, F. (1954). Untersuchungen über die Natur der Winklerschen Burdonen auf Grund neuen experimentellen Materials. *Planta,* **44,** 562–606.

Brabec, F. (1960). Über eine Mesochimäre aus *Solanum nigrum* L. und *Lycopersicon pimpinellifolium* Mill. *Planta,* **55,** 687–707.

Brabec, F. (1965). Pfropfung und Chimären, unter besondere Berücksichtigung der entwicklungsphysiologischen Problematik. In: *Hdb. Pfl. Physiol.,* **15,** Pt. 2. (Ed. W. Ruhland), 388–498. Springer Verlag, Berlin.

Braun, A. (1849). *Betrachtungen über die Erscheinung der Verjüngung in der Natur.* Freiburg i. Br.

Braun, A. (1873). Über *Cytisus adami* und *Syringen*. *Bot. Ztg.*, **31**, 636–8, 647–50, 664–70.

Broertjes, C. and Ballego, M. (1967). Mutation breeding of *Dahlia variabilis*. *Euphytica*, **16**, 171–6.

Broertjes, C., Haccius, B. and Weidlich, S. (1968). Adventitious bud formation on isolated leaves and its significance for mutation breeding. *Euphytica*, **17**, 321–44.

Broertjes, C. and Keen, A. (1980). Adventitious shoots: do they develop from one cell. *Euphytica*, **29**, 73–87.

Broertjes, C. and van Harten, A. M. (1978). *Application of mutation breeding methods in the improvement of vegetatively propagated crops*. Elsevier, Amsterdam and New York.

Buder, J. (1910). Studien an *Laburnum adami*. I. Die Verteilung der Farbstoffe in den Blütenblättern. *Ber. dtsch. bot. Ges.*, **28**, 188–92.

Buder, J. (1911). Studien an *Laburnum adami*. II. Allgemeine anatomische Analyse des Mischlings und seiner Stammpflanzen. *Z. Vererbungsl.*, **5**, 209–84.

Buder, J. (1928). Der Bau des phanerogamen Sproßvegetationspunktes und seine Bedeutung für die Chimärentheorie. *Ber. dtsch. bot. Ges.*, **46**, 20–1.

Burk, L. G. (1975). Clonal and selective propagation of tobacco from leaves. *Plant Science Letters*, **4**, 149–54.

Burk, L. G., Stewart, R. N and Dermen, H. (1964). Histogenesis and genetics of a plastid-controlled chlorophyll variegation in tobacco. *Amer. J. Bot.*, **51**, 713–24.

Bush, S. R., Earle, E. D. and Langhans, R. W. (1976). Plantlets from petal segments, petal epidermis, and shoot tips of the periclinal chimera, *Chrysanthemum morifolium* 'Indianapolis'. *Amer. J. Bot.*, **63**, 729–37.

Butterfass, T. (1979). *Patterns of chloroplast reproduction: a developmental approach to protoplasmic plant anatomy*. Springer-Verlag, Vienna and New York.

Butterfield, H. M. (1926). Thornless blackberry investigations. *California Cultivator*, **67**, 550.

Buvat, R. (1952). Structure, évolution et fonctionnement du méristème apical de quelques dicotyledons. *Ann. Sci. nat. (Bot.) 11ᵉ sér.*, **13**, 202–303.

Buvat, R. (1955). Le méristème apical de la tige. *Ann. Biol.*, **31**, 596–656.

Byatt, J. I., Ferguson, I. K. and Murray, B. G. (1977). Intergeneric hybrids between *Crataegus* L. and *Mespilus* L.: A fresh look at an old problem. *Bot. J. Linn. Soc.*, **74**, 329–44.

Cameron, J. W. and Soost, R. K. (1979). Absence of acidless progeny from crosses of acidless x acidless *Citrus* cultivars. *J. Amer. Soc. Hort. Sci.*, **104**, 220–2.

Cameron, J. W., Soost, R. K. and Olson, E. O. (1964). Chimeral basis for colour in pink and red-fleshed grapefruit. *J. Hered.*, **55**, 23–8.

Caspary, R. (1865). Über Pfropfbastarde. *C. R. Congr. int. Bot. et Hort. (Amsterdam)*, 65–80.

Cassells, A. C. and Minas, G. (1983). Beneficially-infected and chimeral perlargonium: Implications for micropropagation by meristem and explant culture. *Acta Horticulturae*, **131**, 287–97.

Chapot, H. (1964). Des organges sur un citronnier, une chimère? *Cah. Rech. Agron.*, **18**, 117–22.

Chattaway, M. M. and Snow, R. (1929). The genetics of a variegated primrose. *J. Genet.*, **21**, 81–3.

Chittenden, R. J. (1926). Studies in variegation. II. *Hydrangea* and *Pelargonium*. *J. Genet.*, **16**, 43–61.

Chittenden, R. J. (1927). Vegetative segregation. *Bibliogr. genet.*, **3**, 355–442.

Chittenden, R. J. (1928). Ever-sporting races of *Myosotis*. *J. Genet.*, **20**, 123–9.

Chodat, R. (1919). La panachure et les chimères dans le genre *Funkia. C. R. Soc. Phys. Hist. nat. Genève*, **36**, 81–5.

Clark, C. (1930). The origin by mutation of some American potato varieties. *Proc. 17th Ann. Meet. Potato Ass. Amer.*, 117–24.

Clark, C. (1933). Further studies of the origin of russeting in the potato. *Amer. Potato J.*, **10**, 88–91.

Clausen, R. E. and Goodspeed, T. H. (1923). Inheritance in *Nicotiana tabacum*. III. The occurrence of two natural periclinal chimeras. *Genetics*, **8**, 97–105.

Clayberg, C. D. (1963). Delicious apple chimera (half-red, half-yellow). *Fruit Varieties and Hort. Dig.*, **17**, 58.

Clayberg, C. D. (1975). Insect resistance in a graft-induced periclinal chimera of tomato. *Hort Science*, **10**, 13–15.

Clowes, F. A. L. (1957). Chimeras and meristems. *Heredity* **11**, 141–8.

Clowes, F. A. L. (1961). *Apical meristems*. Blackwell Scientific Publications, Oxford.

Collina, F., Mitternpergher, L. and Raddi, P. (1973). La cultivar nettarina 'Angelo Marzocchella' ectochimera monoclamide. *Genetica Agraria*, **27**, 210–6.

Collins, E. J. (1922). Variegation and its inheritance in *Chlorophytum elatum* and *Chlorophytum comosum. J. Genet.*, **12**, 1–17.

Condit, I. J. (1928). Other fig chimeras. *J. Hered.*, **19**, 49–52.

Cook, R. C. (1930). Chimera or accident of development. *J. Hered.*, **21**, 386.

Cornu, A. (1970). Sur l'obtention de mutations somatiques aprés traitements de graines de pétunias. *Ann. Amélior. Plantes*, **20**, 189–214.

Correns, C. (1919). Vererbungsversuche mit buntblättrigen Sippen. II. Vier neue Typen bunter Periklinalchimären. *S.B. preuss. Akad. Wiss.*, 820–57.

Correns, C. (1920). Vererbungsversuche mit buntblättrigen Sippen. III. *Veronica gentianoides albocincta*. IV. Die albomarmorata und albopulverea-Sippen. V. *Mercurialis annus versicolor* und *xantha. S.B. preuss. Akad. Wiss.*, 212–40.

Correns, C. (1922). Vererbungsversuche mit buntblättrigen Sippen. VI–VII. Einige neue Fälle von Albomaculatio. *S.B. preuss. Akad. Wiss.*, 460–86.

Correns, C. (1928). Über nichtmendelnde Vererbung. *Z. Vererbungsl., Suppl.* **1**, 131–68.

Correns, C. (Ed. F. von Wettstein, 1937). *Nicht mendelnde Vererbung. Hb. Vererbungswiss, Bd. II H.* (Eds E. Baur and M. Hartmann). Gebrüder Borntraeger, Berlin.

Cramer, P. J. S. (1907). Kritische Übersicht der bekannten Fälle von Knospenvariation. *Natuurk. Verhand. Holl. Maatsch. D. Wetensch.*, **3e**, 1–474.

Cramer, P. J. S. (1954). Chimeras. *Bibliogr. genet.*, **16**, 193–381.

Crane, M. B (1936). Note on a periclinal chimera in the potato. *J. Genet.*, **32**, 73–7.

Cutter, E. G. (1971). Plant anatomy: experiment and interpretation. *Pt. 2. Organs*. Edward Arnold, London.

Dahlgren, K. V. O. (1953). Die Eigenartigen Vererbungsverhältnisse der Micrantha-Form von *Fragaria vesca. Svensk Bot. Tidskrift*, **47**, 1–15.

Dahlgren, K. V. O. (1959). Pfropf- und Vererbungsversuche mit *Fragaria vesca* F. *micrantha. Svensk Bot. Tidskrift*, **53**, 293–8.

Dahlgren, K. V. O. (1962). Die Lösung des micrantha-Problems bei *Fragaria vesca. Bot. Not.*, **115**, 288–92.

Daker, M. G. (1969). Chromosome numbers of *Pelargonium* species and cultivars. *J. Roy. Hort. Soc.*, **94**, 346–53.

Daniel, L. (1904). Sur un hybride de greffe entre poirer et cognassier. *Rev. gén. Bot.*, **16**, 1–13.

Daniel, L. (1909). Un nouvel hybride de greffe: le néflier de Lagrange. *Rev. bret. Bot.*, **4**, 136–40.

Daniel, L. (1914). L'hybridation asexuelle ou variation spécifique chez les plantes greffées. *Rev. gén. Bot.*, **26**, 305–41.

Daniel, L. (1915). L'hybridation asexuelle ou variation spécifique chez les plantes greffées. *Rev. gén. Bot.*, **27**, 22–9, 33–49.

Daniel, L. and Delpon. J. (1913). Sur un hybride de greffe entre pêcher et amandier. *C.R. Acad. Sci. Paris*, **156**, 2000–2.

Darrow, G. M. (1929). Thornless sports of the young dewberry. *J. Hered.*, **20**, 567–9.

Darrow, G. M. (1931). A productive thornless sport from the evergreen blackberry. *J. Hered.*, **22**, 404–6.

Darrow, G. M., Gilson, R. A., Toenjes, W. E. and Dermen, H. (1948). The nature of giant apple sports and their use in breeding. *J. Hered.*, **39**, 45–51.

Darwin, C. (1868). *The variation of animals and plants under domestication, Pts I and II*. John Murray, London.

De Loose, R. (1979). Radiation induced chimeric rearrangement in flower structure of *Rhododendron simsii* Planch. (*Azalea indica* L.). Use of recurrent irradiation. *Euphytica*, **28**, 105–13.

De Vries, H. (1901–03). *Die Mutationstheorie*. Veit, Leipzig.

Dermen, H. (1945). The mechanism of colchicine-induced cytohistological changes in cranberry. *Amer. J. Bot.*, **32**, 387–94.

Dermen, H. (1947a). Periclinal cytochimeras and histogenesis in cranberry. *Amer. J. Bot.*, **34**, 32–43.

Dermen, H. (1947b). Inducing polyploidy in peach varieties. *J. Hered.*, **38**, 77–82.

Dermen, H. (1947c). Polyploid pears. *J. Hered.*, **38**, 189–92.

Dermen, H. (1947d). Histogenesis of some bud sports and variegations. *Proc. Amer. Soc. Hort. Sci.*, **50**, 51–73.

Dermen, H. (1948). Chimeral apple sports and their propagation through adventitious buds. *J. Hered.*, **39**, 235–42.

Dermen, H. (1950). Pattern reversal in variegated plants. *J. Hered.*, **41**, 325–8.

Dermen, H. (1951a). Tetraploid and diploid adventitious shoots from a giant sport of McIntosh apple. *J. Hered.*, **42**, 144–9.

Dermen, H. (1951b). Ontogeny of tissues in stem and leaf of cytochimeral apples. *Amer. J. Bot.*, **38**, 753–60.

Dermen, H. (1952a). Polyploidy in the apple found seven years after colchicine treatment. *J. Hered.*, **43**, 7–8.

Dermen, H. (1952b). A tetraploid sport of Delicious apple from Idaho. *J. Hered.*, **43**, 8.

Dermen, H. (1953a). The pattern of tetraploidy in the flower and fruit of a cytochimeral apple. *J. Hered.*, **44**, 31–9.

Dermen, H. (1953b). Periclinal cytochimeras and origin of tissues in stem and leaf of peach. *Amer. J. Bot.*, **40**, 154–68.

Dermen, H. (1954). Colchiploidy in grapes. *J. Hered.*, **45**, 159–72.

Dermen, H. (1955). A homogeneous tetraploid shoot from a 2.2.4 type chimeral Winesap apple. *J. Hered.*, **46**, 244.

Dermen, H. (1956). Histogenetic factors in colour and fuzzless peach sports. *J. Hered.*, **47**, 64–76.

Dermen, H. (1960). Nature of plant sports. *Amer. Hort. Mag.*, **39**, 123–73.

Dermen, H. and Bain, H. F. (1941). Periclinal and total polyploidy in cranberries induced by colchicine. *Proc. Amer. Soc. Hort. Sci.*, **38**, 400.

Dermen, H. and Bain, H. F. (1944). A general cytological study of colchicine polyploidy in cranberry. *Amer. J. Bot.*, **31**, 451–63.

Dermen, H. and Darrow, G. M. (1948). A tetraploid sport of McIntosh apples. *J. Hered.*, **39**, 17.

Dermen, H. and Stewart, R. N. (1973). Ontogenetic study of floral organs of peach (*Prunus persica*) utilizing cytochimeral plants. *Amer. J. Bot.*, **60**, 283–91.

Deshayes, A. (1973). Mise en évidence d'une corrélation entre la fréquence de variations somatiques sur feuilles et l'état physiologique d'un mutant chlorophyllien monogénique chez *Nicotiana tabacum* var. Samsun. *Mutation Research*, **17**, 323–34.

Dodds, J. H. and Roberts, L. W. (1982). *Experiments in plant tissue culture.* Cambridge University Press, Cambridge.

Dommergues, P. and Gillot, J. (1973). Obtention de clones génétiquement homogènes dans toutes leur couches ontogénétiques a partir d'une chimère d'oeillet américain. *Ann. Amélior. Plantes*, **23**, 83–95.

Doodeman, M. and Bianchi, F. (1985). Genetic analysis of the instability in *Petunia* hybrids. 3. Periclinal chimeras resulting from frequent mutations of unstable alleles. *Theor. Appl. Genet.*, **69**, 297–304.

Dormer, K. J. (1980). *Fundamental tissue geometry for biologists.* Cambridge University Press, Cambridge.

Dorst, J. C. (1952). Two remarkable bud-sports in the potato variety Rode Star. *Euphytica*, **1**, 184–6.

Drain, B. D. (1932). Field studies of bud sports in Michigan tree fruits. *Michigan Agric. Exp. Sta. Tech. Bull.*, **130**, 1–48.

Duckett, J. G. and Toth, R. (1977). Giant mitochondria in a periclinal chimera, *Ficus elastica* Roxb. forma *variegata*. *Ann. Bot.*, **41**, 903–12.

Dulieu, H. (1965). Reáctions morphologiques des plantules de *Nicotiana tabacum* L. et de *Nicotiana glutinosa* L. au traitement des graines par le méthane sulfonate d'éthyle. Conséquences sur l'apparition de mutations chlorophylliennes somatiques. *Ann. Amélior. Plantes*, **15**, 359–72.

Dulieu, H. (1967a). Sur les différents types de mutations extra-nucléairies induites par le méthane sulfonate d'éthyle chez *Nicotiana tabacum* L. *Mutation Research*, **4**, 177–89.

Dulieu, H. (1967b). Étude de la stabilité d'une déficience chlorophyllienne induite chez le tabac par traitement au méthane sulfonate d'éthyle. *Ann. Amélior. Plantes*, **17**, 339–55.

Dulieu, H. (1968). Emploi des chimères chlorophylliennes pour l'étude de l'ontogénie foliaire. *Bull. Sci. Bourgogne*, **25**, 1–60.

Dulieu,, H. (1969). *Mutations somatiques chlorophylliennes induites et ontogénie caulinaire.* Part of a thesis submitted for de Doctorat d'Etat des Sciences naturelles, Dijon, France.

Dulieu, H. (1970). Les mutations somatiques induites et l'ontogénie de la pousse feuillée. *Ann. Amélior. Plantes*, **20**, 27–44.

Einset, J. (1948). The occurrence of spontaneous triploids and tetraploids in apples. *Proc. Amer. Soc. Hort. Sci.*, **51**, 61–3.

Einset, J. (1950). The Wrixparent, a 'tetraploid' apple. *Proc. Amer. Soc. Hort. Sci.*, **55**, 262.

Einset, J. (1952). Spontaneous polyploidy in cultivated apples. *Proc. Amer. Soc. Hort. Sci.*, **59**, 291–302.

Einset, J., Blaser, H. W. amd Imhofe, B. (1946). A chromosomal chimera of the Northern Spy apple. *J. Hered.*, **37**, 265–6.

Einset, J., Blaser, H. W. and Imhofe, B. (1947). Chimeral sports of apples. *J. Hered.*, **38**, 371–6.

Einset, J. and Imhofe, B. (1949). Chromosome numbers of apple varieties and sports. *Proc. Amer. Soc. Hort. Sci.*, **53**, 197–201.

Einset, J. and Lamb, B. (1951a). Chimeral sports of grapes: Alleged tetraploid varieties have diploid 'skin'. *J. Hered.*, **42**, 158–62.

Einset, J. and Lamb, B. (1951b). Chromosome numbers of apple varieties and sports III. *Proc. Amer. Soc. Hort. Sci.*, **58**, 103–8.

Einset, J. and Pratt, C. (1954). 'Giant' sports of grapes. *Proc. Amer. Soc. Hort. Sci.*, **63**, 251–6.

Einset, J. and Pratt, C. (1959). Spontaneous and induced apple sports with misshapen fruit. *Proc. Amer. Soc. Hort. Sci.*, **73**, 1–8.

Emsweller, S. L. (1947). The utilization of induced polyploidy in Easter Lily breeding. *Proc. Amer. Soc. Hort. Sci.*, **49**, 379–84.

Emsweller, S. L. (1949). Colchicine-induced polyploidy in *Lilium longiflorum*. *Amer. J. Bot.*, **36**, 135–44.

Emsweller, S. L. and Stewart, R. N. (1951). Diploid and tetraploid pollen mother cells in lily chimeras. *Proc. Amer. Soc. Hort. Sci.*, **57**, 414–8.

Esau, K. (1965). *Plant anatomy*, 2nd edition. John Wiley, New York.

Farestveit, B. (1969). Flower colour chimeras in glasshouse carnations, *Dianthus caryophyllus* L. *Arsskrift K. Veterinaer-og Landbohkojskole Kobenhaven*, 19–33.

Fischer, E. (1912). Die Empfänglichkeit von Pfropfreisern und Chimären für Uredineen. *Mycol. Zbl.*, **1**, 195–9.

Fisher, J. (1982). *The origins of garden plants.* Constable, London.

Focke, W. O. (1877). Miszellen I, Variationen an gescheckten Hülsen. *Abh. d. Naturw. Ver. Bremen*, **5**, 401.

Foster, A. S. (1938). Structure and growth of the shoot apex in *Ginkgo biloba*. *Bull, Torrey bot. Club*, **65**, 531–56.

Foyle, H. W. and Dermen, H. (1969). Genetic and chimeral constitution of three leaf variegations in the peach. *J. Hered.*, **60**, 323–8.

Frandsen, N. O. (1967). Chromosomenverdoppelung und Chimärenbildung nach Colchicin-behandlung haploider Kartoffelsamen. *Eur. Potato J.*, **19**, 1–15.

Frost, H. B. (1926). Polyembryony, heterozygosis and chimeras in *Citrus*. *Hilgardia*, **1**, 365–401.

Frost, H. B. and Krug, C. A. (1942). Diploid-tetraploid periclinal chimeras as bud variants in *Citrus*. *Genetics*, **27**, 619–34.

Fry, B. O (1963). Production of tetraploid muscadine (*V. rotundifolia*) grapes by gamma radiation. *Proc. Amer. Soc. Hort. Sci.*, **83**, 388–94.

Fuchs, C. A. (1898). Untersuchungen über *C. adami*. *Sitzs. Kais. Akad. Wiss. Wien*, **107**, 1273–92.

Fucik, V. (1960). The seed progeny of Solanaceae chimeras. *Biol. Plant.*, **2**, 216–22.

Funaoka, S. (1924). Beiträge zur Kenntnis der Anatomie panaschierter Blätter. *Biol. Zbl.*, **44**, 343–84.

Gardner, V. R. (1944). A study of the sweet-and-sour apple chimera and its clonal significance. *J. Agric. Res.*, **68**, 383–94.

Gardner, V. R., Crist, J. W. and Gibson, R. E. (1933). Somatic segregation in a sectorial chimera of the Bartlett pear. *J. Agric. Res.*, **46**, 1047–57.

Gardner, V. R., Toenjes, W. and Giefel, M. (1948). Segregation in a radially unsymmetrical sport of the Canada Red apple. *J. Agric. Res.*, **76**, 241–55.

Gifford, E. M. Jr., and Corson, G. E. Jr., (1971). The shoot apex in seed plants. *Bot. Rev.*, **37**, 143–229.

Gottschalk, W. and Wolff, G. (1983). *Induced mutations in plant breeding.* Springer-Verlag, Berlin.

Greis, H. (1940). Vergleichende physiologische Untersuchungen an diploiden und tetraploiden Gersten. *Züchter*, **12**, 62–73.

Grimbly, P. E. (1977). Tomato silvering, its anatomy and chimeral structure. *J. Hort. Sci.*, **52**, 469–73.

Grimbly, P. E. and Thomas, B. J. (1977). Silvering, a disorder of the tomato. *J.*

Hort. Sci., **52**, 49–57.

Guillaumin, A. (1949). A propos des chimères. *Ann. Sci. nat. Bot.*, *11ᵉ serie*, **10**, 1–19.

Günther, E. (1954). Untersuchungen an einer Solanaceen-Chimäre (*Lycopersicon esculentum* + *Solanum nigrum*). *Wiss. Z. Univ. Greifswald*, **3**, 161–71.

Günther, E. (1957a). Untersuchungen an Monoektochimären zwischen *Lycopersicon esculentum* Mill. und *Solanum nigrum* L. *Biol. Zbl.*, **76**, 343–51.

Günther, E. (1957b). Die Nachkommenschaft von Solanaceen-Chimären. (1. Mitteilung). *Flora*, **144**, 497–517.

Günther, E. (1961). Durch Chimärenbildung verursachte Aufhebung der Selbstinkompatibilität von *Lycopersicon peruvianum* (L) Mill. *Ber. dtsch. bot. Ges.*, **74**, 333–6.

Günther, E. (1962). Die Nachkommenschaft von Solanaceen-Chimären. (2. Mitteilung). *Flora*, **152**, 196–226.

Guttenberg, H. von (1961). *Grundzüge der Histogenese höherer Pflanzen. II. Gymnospermen.* Gebrüder Borntraeger, Berlin.

Haberlandt, G. (1926). Über den Blattbau der Crataegomespili von Bronvaux und ihrer Eltern. *Sitzs. preuss. Akad. Wiss.*, *Physik.-math. Kl.*, **17**, 170–208.

Haberlandt, G. (1927). Sind die Crataegomespili von Bronvaux Verschmel-zungspfropfbastarde der Periklinalchimären. *Biol. Zbl.*, **47**, 129–51.

Haberlandt, G. (1930). Das Wesen der Crataegomespili. *Sitzs. preuss. Akad. Wiss.*, *Physik.-math. Kl.*, **20**, 374–94.

Haberlandt, G. (1931). Was sind die Crataegomespili? *Biol. Zbl.*, **51**, 253–9.

Haberlandt, G. (1934a). Blattepidermis und Palisadengewebe der Crataegomespili und ihrer Eltern. *Sitzs. preuss. Akad. Wiss.*, *Physik.-math. Kl.* (1934), 178–90.

Haberlandt, G. (1934b). Über die Sonnen- und Schattenblätter der Crataegomespili und ihrer Eltern. *Sitzs. preuss. Akad. Wiss.*, *Physik.-math. Kl.* (1934), 365–76.

Haberlandt, G. (1935). Beiträge zum Crataegomespilus-Problem. *Sitzs. preuss. Akad. Wiss.*, *Physik.-math. Kl.* (1935). 480–99.

Haberlandt, G. (1941). Über das Wesen der morphogenen Substanzen. *Sitzs. preuss. Akad. Wiss.*, *Physik.-math. Kl.* (1941), 3–10.

Hagemann, R. (1964). *Plasmatische Vererbung.* Veb Gustav Fischer Verlag, Jena.

Hanstein, J. (1868). Die Scheitelzellgruppe in Vegetationspunkt der Phanerogamen. *Festschr. Neiderrhein. Gesell, Natur- und Heilkunde*, 109–34.

Harlan, J. R. and de Wet, J. M. J. (1975). The origins of polyploidy. *Bot. Rev.*, **41**, 361–90.

Haskell, G. (1965). Biochemical differences in colour sectors of a chimeral orange fruit. *J. Hered.*, **56**, 35–7.

Heichel, G. H. and Anagnostakis, S. L. (1978). Stomatal response to light of *Solanum pennellii*, *Lycopersicon esculentum*, and a graft-induced chimera. *Plant Physiol.*, **62**, 387-90.

Heiken, A. (1958). Aberrant types in the potato. *Acta Agric. J. cand.*, **8**, 319–58.

Heiken, A. and Ewertson, G. (1962). The chimerical structure of a somatic *Solanum* mutant revealed by ionizing irradiation. *Genetica*, **33**, 88–94.

Heiken, A., Ewertson, G. and Carlström, L. (1963). Studies on a somatic subdivided-leaf mutant in *Solanum tuberosum*. *Radiat. Bot.*, **3**, 145–53.

Hejnowicz, Z. (1956). The first periclinal chimera among Gymnosperms. *Acta Soc. Bot. Polon.*, **25**, 181–202. (Polish)

Hejnowicz, Z. (1959). Eversporting periclinal chimeras. *Recent Advances in Botany*, **2**, 1446–8.

Hénon, J. L. (1839). Note sur le Cytise Labour, Adam et purpre. *Ann. Sci. Phys. Nat. Agr. Ind. Lyon*, **2**, 375–7.

Hermsen, J. G. T. and de Boer, A. J. E. (1971). The effect of colchicine treatment on *Solanum acaule* and *S. bulbocastanum:* A complete analysis of ploidy chimeras in *S. bulbocastanum. Euphytica*, **20**, 171–80.

Heuer, W. (1910). Pfropfbastarde zwischen Solanum-Arten. *Gartenflora*, **59**, 434–8.

Hildebrand, F. (1908). Über Sämlinge von *C. adami. Ber. dtsch. bot. Ges.*, **26a**, 590–5.

Hjelmqvist, H. (1937). Ein paar neue Crataegomespili. *Hereditas*, **22**, 376–94.

Hjelmqvist, H. (1947). Eine Periklinalchimäre in der Gattung Syringa. *Hereditas*, **33**, 367–76.

Honing, J. A. (1927). Erblichkeitsuntersuchungen an Tabak. *Genetica*, **9**, 1–18.

Hosticka, L. P. and Hanson, M. R. (1984). Induction of plastid mutations in tomatoes by nitrosomethylurea. *J. Hered.*, **75**, 242–6.

Howard, H. W. (1958). Transformation of a monochlamydius into a dichlamydius chimaera by X-ray treatment. *Nature*, **182**, 1620.

Howard, H. W. (1959). Experiments with a potato periclinal chimaera. *Genetica*, **30**, 278–91.

Howard, H. W. (1961a). The production of hexaploid *Solanum* × *Juzepczukii. Euphytica*, **10**, 95–100.

Howard, H. W. (1961b). Mericlinal chimeras in the potato variety Gladstone. *New Phytol.*, **60**, 388–92.

Howard, H. W. (1962). Experiments with potatoes on the effect of the pigment-restricting gene, M. *Heredity*, **17**, 145–56.

Howard, H. W. (1964a). The experimental production of buds on the roots of potatoes. *Nature*, **203**, 1303–4.

Howard, H. W.(1964b). The use of X-rays in investigating potato chimeras. *Radiat. Bot.*, **4**, 361–71.

Howard, H. W. (1965). The chimerical nature of the entire-leaf variant in the potato variety 'Majestic'. *Nature*, **208**, 197.

Howard, H. W. (1966a). Recombination value for genes E and M in potatoes. *Heredity*, **21**, 313–5.

Howard, H. W. (1966b). A sectorial entire-pinnate leaf chimera in the potato variety Majestic. *New Phytol.*, **65**, 284–7.

Howard, H. W. (1966c). Periclinal nature of the ivy-leaf sport in potatoes. *Nature*, **209**, 108–9.

Howard, H. W. (1967a). Differentiation in potatoes: hidden-spotted and spectacled. *Heredity*, **22**, 57–64.

Howard, H. W. (1967b). Further experiments on the use of X-rays and other methods in investigating potato chimeras. *Radiat. Bot.*, **7**, 389-99.

Howard, H. W. (1967c). The chimerical nature of a potato wilding. *Pl. Path.*, **16**, 89–92.

Howard, H. W. (1969a). A full analysis of a potato chimera. *Genetica*, **40**, 233–41.

Howard, H. W. (1969b). The chimeral nature of a feathery wilding.*Eur. Potato J.*, **12**, 67–9.

Howard, H. W. (1970a). *Genetics of the potato, Solanum tuberosum.* Logos Press, London.

Howard, H. W. (1970b). The eye-excision method of investigating potato chimeras. *Potato Res.*, **13**, 220–2.

Howard, H. W. (1971a). The stability of L I-mutant periclinal potato chimeras. *Potato Res.*, **14**, 91–3.

Howard, H. W. (1971b). A sectorial green-yellow leaf chimera in the potato. *New Phytol.*, **70**, 873–8.

Howard, H. W., Wainwright, J. and Fuller, J. M. (1963). The number of independent layers at the stem apex in potatoes. *Genetica*, **34**, 113–20.

Hughes, H. M. (1963). A study of two black current chimeras. *J. Hort. Sci.*, **38.**, 286–96.

Humihiko, H. (1941). Intergeneric hybridization in Cichorieae. V. Variation in karyotypes and fertility of *Crepidiastrixeris denticulatum platyphylla*. *Cytologia*, **11,** 338–52.

Hutchins, A. E. and Younger, V. B. (1952). Maternal inheritance of a colour variation (chimaera) in the squash, *Cucurbita maxima* Duch. *Proc. Amer. Soc. Hort. Sci.*, **60,** 370–8.

Ikeno, S. (1934). Studien über die mutative Entstehung eines hochmutables Genes bei einer parthenogenetischen Pflanzenart. *Z. Vererbungsl.*, **68,** 517–42.

Imai, Y. (1928). A consideration of variegation. *Genetics*, **13,** 544–62.

Imai, Y. (1934). On the mutable genes of *Pharbitis nil*, with special reference to their bearing on the mechanism of bud-variation. *J. Coll. Agric. Tokyo*, **12,** 479–523.

Imai, Y. (1935a). Variegated flowers and their derivatives by bud variation. *J. Genet.*, **30,** 1–13.

Imai, Y. (1935b). The structure of albomarginata and medioalbinata forms. *J. Genet.*, **31,** 53–65.

Imai, Y. (1935c). The mechanism of bud variation. *Amer. Nat.*, **69,** 587–95.

Imai, Y. (1936a). Chlorophyll variegations due to mutable genes and plastids. *Z. Vererbungsl.*, **71,** 61–83.

Imai, Y. (1936b). Geno- and plasmotypes of variegated Pelargoniums. *J. Genet.*, **33,** 169–95.

Imai, Y. (1937). Bud variation in a flaked strain of *Rhododendron obtusum*. *J. Coll. Agric. Tokyo*, **14,** 93–8.

Jamieson, A. P. and Willmer, C. M. (1984). Functional stomata in a variegated leaf chimera of *Pelargonium zonale* L. without guard cell chloroplasts. *J. Exp. Bot.*, **35,** 1053–9.

Johnson, M. A. (1951). The shoot apex in gymnosperms. *Phytomorph.*, **1,** 188–204.

Johnson, R. T. (1980). Gamma irradiation and in vitro induced separation of chimeral genotypes in carnation. *Hort Science*, **15,** 605–6.

Johnsson, H. (1950). On the C_0 and C_1 generations in *Alnus glutinosa*. *Hereditas*, **36,** 205–19.

Jörgensen, C. A. (1927). A periclinal tomato-potato chimera. *Hereditas*, **10,** 293–302.

Jörgensen, C. A. and Crane, M. B. (1927). Formation and morphology of *Solanum* chimaeras. *J. Genet.*, **18,** 247–73.

Jouin, E. (1899). Peut-on obtenir des hybrides par le greffage? *Jardin (Paris)*, **20,** 22–4.

Junker, G. and Mayer, W. (1974). Die Bedeutung der Epidermis für licht und temperatur- induzierte Phasenverschiebungen circadianer Laubblätt-bewegungen. *Planta*, **121,** 27–37.

Kameya, T. (1975). Culture of protoplasts from chimeral plant tissue of nature. *Jap. J. Genet.*, **50,** 417–20.

Kasperbauer, M. J., Sutton, T. G., Anderson, R. A. and Gupton, C. L. (1981). Tissue culture of plants from a chimeral mutation of tobacco. *Crop Science*, **21,** 588–90.

Katagiri, K. (1976). Radiation damage and induced tetraploidy in mulberry (*Morus alba* L.). *Environmental and Exp. Bot.*, **16,** 119–30.

Katagiri, K. and Nakajima, K. (1982). Tetraploid induction by gamma-ray irradiation in mulberry. In: *Induced mutations in vegetatively propagated plants II*. Coimbatore, India, 11–15 February 1980. IAEA, Vienna.

Keeble, F. and Armstrong, E. F. (1912). The oxydases of *Cytisus adami*. *Proc. Roy. Soc. London, B.*, **85,** 460–5.

Kerns, K. R. and Collins, J. L. (1947). Chimeras in pineapple. Colchicine induced tetraploids and diploid tetraploids in the Cayenne variety. *J. Hered.*, **38**, 323–30.

Kirk, J. T. O. and Tilney-Bassett, R. A. E. (1978). *The Plastids: Their chemistry, structure, growth and inheritance*. 2nd edition, Elsevier/North-Holland, Amsterdam.

Klebahn, H. (1918). Impfversuche mit Pfropfbastarden. *Flora*, **111**, 418–30.

Klopfer, K. (1965a). Über den Nachweis von drei selbständigen Schichten im Sproßscheitel der Kartoffel. *Z. Pflanzenzücht.*, **53**, 67–87.

Klopfer, K. (1965b). Erfolgreiche experimentelle Entmischungen und Umlagerungen periclinalchimärischer Kartoffelklone. *Züchter*, **35**, 201–14.

Klopfer, K. (1965c). Histogenetische Untersuchungen am Sproßscheitel der Kartoffel. *Flora*, **156**, 50–77.

Koehne, E. (1901). Zwei Pfropfbastarde von *Crataegus monogyna* und *Mespilus germanica*. *Gartenflora*, **50**, 628–33.

Kosichenko, N. E. and Petrov, S. A. (1975). Anatomic and physiological characteristics of the leaves of chimeral poplar Kazakhstanskii 272 and grafting components. *Bot. Zh. (Leningr)*. **60**, 1331–5. (Russian)

Kotila, J. E. (1929). Some bud mutations in the potato. *Amer. Potato J.*, **6**, 131–5.

Krantz, F. A. (1951). Potato breeding in the United States. *Z. Pflanzenzücht.*, **29**, 388–93.

Krantz, F. A and Tolaas, A. G. (1939). The Red Warba potato. *Amer. Potato J.*, **16**, 185–90.

Krenke, N. P. (1929). Chimären zwischen *Saracha umbellata* Dan. und *Solanum lycopersicum* L. *Proc. USSR Congr. Gen. (Leningr.) Bd.*, **2**, 319–432.

Krenke, N. P. (1933). *Wundkompensation, Transplantation und Chimären bei Pflanzen*. Springer-Verlag, Berlin.

Krüger, M. (1931). Vergleichend-entwicklungsgeschichtliche Untersuchungen an den Fruchtkoten und Früchten zweier *Solanum*-Chimären und ihrer Elternarten. *Planta*, **17**, 372–436.

Kümmler, A. (1922). Über die Funktion der Spaltöffnungen weissbunter Blätter, *Jb. Wiss. Bot.*, **61**, 610–69.

Küster, E. (1919). Über weissrandige Blätter und andere Formen der Buntblättrigkeit. *Biol. Zbl.*, **39**, 212–51.

Küster, E. (1926). Beiträge zur Kenntnis der panáschierten Gehölze. XIV–XVII. *Mitteil. d. D. dendrol.Ges.*, **36**, 258–71.

Küster, E. (1927). Anatomie des panaschierten Blattes. In: *Handbuch der Pflanzenanatomie. II Abteilung, 2. Teil: Pteridophyten und Anthophyten, Bd VIII* (Ed. K. Linsbauer), 1–68. Gebrüder Borntraeger, Berlin.

Küster, E. (1928). Beiträge zur Kenntnis der panaschierten Gehölze. XVIII–XXII. *Mitteil. d. D. dendrol. Ges.*, **37**, 258–67.

Küster, E. (1929). Beiträge zur Kenntnis der panaschierten Gehölze. XXIII–XXVII. *Mitteil. d. D. dendrol, Ges.*, **38**, 347–56.

Küster, E. (1932). Beiträge zur Kenntnis der panaschierten Gehölze. XXXIV. Randpanaschierung bei *Prunus cerasifera Pissartii*. *Mitteil. d. D. dendrol. Ges.*, **44**, 391–2.

Küster, E. (1934). Beiträge zur Morphologie der panaschierten Gewächse. *Biol. Zbl.*, **54**, 89–95.

Küster, E. (1937). Beiträge zur Kenntnis der panaschierten Gehölze. *Mitteil. d. D. dendrol. Ges.* **49**, 70–9.

Lacey, C. N. D. (1982). The stability of induced compact mutant clones of Bramley's seedling apple. *Euphytica*, **31**, 451–9.

Lakon, G. (1921). Die Weissrandpanaschierung von *Acer negundo*. *Z. Vererbungsl.*, **26**, 271–84.

Lamprecht, H. and Svensson, V. (1949). Zwei Chimären von *Daucus carota* L. sowie allgemeines über Art und Enstehung von Chimären. *Agri. Hort. Genet.*, **7**, 96–111.

Lange, F. (1927). Vergleichende Untersuchungen über die Blattentwicklung einiger *Solanum*-Chimären und ihrer Elternarten. *Planta*, **3**, 181–282.

Lange, F. (1933). Über die Blattentwicklung der Crataegomespili von Bronvaux und ihrer Elternarten. *Planta*, **20**, 1–44.

Langton, F. A. (1980). Chimerical structure and carotenoid inheritance in *Chrysanthemum morifolium* (Ramat.). *Euphytica*, **29**, 807–12.

Laubert, R. (1901). Anatomische und morphologische Studien am Bastard *Laburnum adami. Beih. Bot. Zbl.*, **10**, 144–65.

Lesley, J. W. (1929). A few-seeded bud sport of tomato. *J. Hered.*, **20**, 531–3.

Lieske, R. (1921). Pfropfversuche. IV. Untersuchungen über die Reizleitung der Mimosen. *Ber dtsch. bot. Ges.*, **39**, 348–50.

Lieske, R. (1927). Demonstration. *Ber. dtsch. bot. Ges.*, **45**, 142.

Lindemuth, H. (1907). Studien über die sogenannte Panaschure und über begleitende Erscheinungen. *Landw. Jahrb.*, **36**, 807–63.

Lindgren, D., Eriksson, G. and Sulovska, K. (1970). The size and appearance of the mutated sector in barley spikes. *Hereditas*, **65**, 107–32.

Linsbauer, K. (1931). Histologische Notizen. I. A. Panaschierte Opuntien. B. Über einige Spaltoffnungsanomalien. *Ber. dtsch. bot. Ges.*, **49**, 64–76.

MacArthur, J. W. (1928). A spontaneous tomato chimera. *J. Hered.*, **19**, 331–4.

Macfarlane, M. J. (1892). A comparison of the minute structure of plant hybrids with that of their parents, and its bearing on biological problems. *Trans. Roy. Soc. Edinb.*, **37**, 203–86.

Maliga, P., Kiss, Z. R., Nagy, A. H. and Lazar, G. (1978). Genetic instability in somatic hybrids of *Nicotiana tabacum* and *Nicotiana knightiana. Mol. Gen. Genet.*, **163**, 145–51.

Manenti, G. (1975). The structure of variegated leaves of *Acer negundo* L. A light and electron microscope study. *Israel J. Bot.*, **24**, 61–70.

Marcotrigiano, M. and Gouin, F. R. (1984a). Experimentally synthesized plant chimeras 1. *In vitro* recovery of *Nicotiana tabacum* L. chimeras from mixed callus cultures. *Ann. Bot.*, **54**, 503–11.

Marcotrigiano, M. and Gouin, F. R. (1984b). Experimentally synthesized plant chimeras 2. A comparison of *in vitro* and *in vivo* techniques for the production of interspecific *Nicotiana* chimeras. *Ann. Bot.*, **54**, 513–21.

Marks. G. E. (1953). Genetical studies in pears. VI. Giant bud sports. *J. Hort. Sci.*, **28**, 141–4.

Mason, S. C. (1930). A sectorial mutation of a Deglet Noor date palm. *J. Hered.*, **21**, 157–63.

Massey, K. (1928). The development of the leaves in certain periclinally variegated plants. *J. Genet.*, **19**, 357–72.

Maurizio, A. M. (1927). Zur Biologie und Systematik der Pomaceen bewohnenden Podosphaeren. Mit Berücksichtigung der Frage der Empfänglichkeit der Pomaceenpfropfbastarde für parasitische Pilze. *Zbl. Bakt. Abt.*, **72**, 129–48.

Mayer, W., Moser, I. and Bünning, E. (1973). Die Epidermis als Ort der Licht–perzeption duer circadiane Laubblattbewegungen und photoperiodische Induktionen. *Z. Pflanzenphysiol.*, **70**, 66–73.

Mayer-Alberti, M. (1924). Vergleichende Untersuchungen über den Blattbau einiger *Solanum* Pfropfbastarde. *Mitteil. Inst. Allg. Bot. Hamburg*, **6**, 1–32.

McIntosh, T. P. (1927). *The potato. Its history, varieties, culture and diseases.* Oliver and Boyd, Edinburgh.

McIntosh, T. P. (1945). Variations in potato varieties. *Scot. J. Agric.*, **25**, 125–32.

McPheeters, K. and Skirvin, R. M. (1983). Histogenic layer manipulation in chimeral 'Thornless Evergreen' trailing blackberry. *Euphytica*, 32, 351–60.

Mehlquist, G. A. L. and Geissman, T. A. (1947). Inheritance in the carnation *(Dianthus caryophyllus)*. III. Inheritance of flower colour. *Ann. Missouri Bot. Garden*, 34, 39–74.

Mehlquist, G. A. L. and Sagawa, Y. (1964). The effect of gamma radiation on carnations. *Proc. Int. Hort. Congr.*, 16, 10–18.

Melchers, G. (1960). Haploide Blütenpflanzen als Material der Mutationszüchtung. Beispiele: Blattfarbmutanten und mutatio wettsteinii von *Antirrhinum majus*. *Züchter*, 30, 129–34.

Melchers, G. and Bergmann, L. (1958). Untersuchungen an Kulturen von hapoiden Geweben von *Antirrhinum majus*. *Ber. dtsch. bot. Ges.*, 71, 459–73.

Melchers, G. and Labib, G. (1970). Die Bedeutung haploider höheren pflanzen für Pflanzenphysiologie und Pflanzenzüchtung. *Ber. dtsch. bot. Ges.*, 83, 129–50.

Meyer, J. (1915). Die Crataegomespili von Bronvaux. *Z. Vererbungsl.*, 13, 193–233.

Michaelis, P. (1957). Genetische, entwicklungsgeschichtliche und cytologische Untersuchungen zur Plasmavererbung. II. Mitteilung. Über eine Plastidenmutation mit intracellularer Wechselwirkung der Plastiden, zugleich ein Beitrag zur Methodik der Plasmonanalyse und zur Entwicklunggeschichte von *Epilobium*. *Planta*, 50, 60–106.

Miedema, P. (1967). Induction of adventitious buds on roots of the potato. *Euphytica*, 16, 163–6.

Miller, P. D., Vaughn, K. C. and Wilson, K. G. (1980). Induction, ultrastructure, isolation, and tissue culture of chlorophyll mutants in carrot. *In Vitro*, 16, 823–8.

Miller, P. D. Vaughn, K. C. and Wilson, K. G. (1984). Ethyl methane-sulfonate-induced chloroplast mutagenesis in crops. *J. Hered.*, 75, 86–92.

Miyake, K. and Imai, Y. (1934). A chimeral strain with variegated flowers in *Chrysanthemum sinense*. *Z. Vererbungsl.*, 68, 300–2.

Moh, C. C. (1961). Does a coffee plant develop from one initial cell in the shoot apex of an embryo? *Radiat. Bot.*, 1, 97–9.

Morren, E. (1871). Notice sur le *Cytise × purpureo-laburnum, Cytisus adami* Poit. *Belg. Hort. Gand.*, 21, 225–37.

Nati, P. (1674). *Florentina phytologica: Observatio de malo Limonia citrata-aurantia Florentinae vulgo La Bizzaria*. Florenz.

Neilson-Jones, W. (1934). *Plant chimeras and graft hybrids*. Methuen, London.

Neilson-Jones, W. (1937). Chimeras: a summary and some special aspects. *Bot. Rev.*, 3, 545–62.

Neilson-Jones, W. (1969). *Plant chimeras*. 2nd edition, Methuen, London.

Nemec, B. (1910). *Das Problem der Befruchtungsvorgänge und andere zytologische Fragen*. Gebrüder Borntraeger, Berlin.

Nevers, P., Shepherd, N. S. and Saedler, H. (1986). Plant transposable elements. *Advances Bot. Res.*, 12, 103–203.

Newcomer, E. H. (1941). A colchicine-induced tetraploid cabbage. *Amer. Nat.*, 75, 620.

Newman, I. V. (1961). Pattern in the meristem of vascular plants. II. A review of shoot apical meristems of gymnosperms, with comments on apical biology and taxonomy, and a statement of some fundamental concepts. *Proc. Linn. Soc. (New So. Wales)*, 86, 9–59.

Newman, I. V. (1965). Pattern in the meristem of vascular plants. III. Pursuing the patterns in the apical meristem where no cell is a permanent cell. *J. Linn Soc. (Bot.)*, 59, 185–214.

Nilsson, F. and Goldschmidt, E. (1962). Cytochimeras in Ribes. *Bot. Notiser.*, 115, 137–46.

Nishiura, M. and Iwamasa, M. (1969). Chimeral basis for vegetative reversion in the Suzuki Wase, an early maturing mutant of Satsuma mandarin. *Bull. Hort. Res. Stå. Ser. B. (Okitsu)*, **9**, 1–8.

Nishiura, M. and Iwamasa, M. (1970). Reversion of fruit colour in nucellar seedlings from the Dobashibeni Unshu, a red colour mutant of the satsuma mandarin. *Bull. Hort. Res. Sta. Ser. B. (Okitsu)*, **10**, 1–5.

Noack, K. L. (1922). Entwicklungsmechanische Studien an panaschierten Pelargonien. Zugleich ein Beitrag zur Theorie der Periklinalchimären. *Jb. Wiss. Bot.*, **61**, 459–534.

Noack, K. L. (1930). Untersuchungen an *Pelargonium* 'Freak of Nature'. *Z. Bot.*, **23**, 309–327.

Noack, K. L. (1932). Über Hypericum-Kreuzungen. II. Rassen- und Artkreuzungen mit einem buntblättrigen *Hypericum acutum. Z. Vererbungsl.*, **63**, 232–55.

Noirot, M. (1978). Polyploidisation de caféiers par la colchicine. Adaptation de la technique sur bourgeons axillaires aux conditions de Madagascar. Mise en évidence de chimères périclines stables. *Café Cacao Thé*, **22**, 187—94.

Noll, F. (1905). Die Pfropfbastarde von Bronvaux. *Sitzb. Niederrhein Ges. f. Nat. u. Heilk.*, A 20 and A 53.

Noll, F. (1907). Neue Beobachtungen an Laburnum adami. *Sitzb. Neiderrhein Ges. f. Nat. u. Heilk.*, A 38.

Ohri, D. amd Khoshoo, T. N. (1982). Cytogenetics of cultivated Bougainvilleas. X. Nuclear DNA content. *Z. Pflanzenzücht.*, **88**, 168–73.

Olson, E. O. Cameron, J. W. and Soost, K. (1966). The Burgundy sport: Further evidence of the chimeral nature of pigmented grapefruits. *HortScience*, **1**, 57–8.

Opatrny, Z. and Landa, Z. (1974). Regeneration of chlorophyll chimeras from leaf explants of *Nicotiana tabacum* L. *Biologia Plantarum*, **16**, 312–15.

Penzig, O. (1887). La Bizzaria. *Bull. Soc. Toscana ortic.*, **12**, 78–87.

Péreau-Leroy, P. (1969). Effect de l'irradiation gamma sur une chimère complexe d'oeillet sim. In: *Induced mutations in plants. Proc. Symp. FAO/IAEA*, Vienna, 337–44.

Péreau-Leroy, P. (1974). Genetic interaction between the tissues of carnation petals as periclinal chimeras. *Radiat. Bot.*, **14**, 109–16.

Pierik, R. L. M. and Steegmans, H. H. M. (1983). Vegetative propagation of a chimeral *Yucca elephantipes* Regel *in vitro. Scientia Hortic.*, **21**, 267–72.

Pohlheim, F. (1968). *Thuja gigantea gracilis* Beissn. – ein Haploid unter den Gymnospermen. *Biol. Rdsch.*, **6**, 84–6.

Pohlheim, F. (1969a). Untersuchungen über Chloroplastenzahlen in Schliesszellen, Schliesszellenlängen und Kerngrössen in Epidermiszellen bei *Laburnocytisus-Chimären und ihre Rückschlagen. Wiss. Z. Päd. Hochsch. Potsdam*, **13**, 157–65.

Pohlheim, F. (1969b). Über Unterschiede in der Beteiligung des 'Dermatogens' an der Mesophyllbildung bei *Buxus sempervirens* argenteo-marginata hort. *Wiss. Z. Päd. Hochsch. Potsdam*, **13**, 167–76.

Pohlheim, F. (1970a). *Prunus pissardi* 'Hessei' – eine Trichimäre. *Flora*, **159**, 435–49.

Pohlheim, F. (1970b). Triploidie bei *Thuja plicata excelsa* Timm. *Biol. Rdsch.*, **8**, 402–3.

Pohlheim, F. (1971a). Untersuchungen zur Sprossvariation der Cupressaceae. 1. Nachweis immerspaltender Periklinalchimären. *Flora*, **160**, 264–93.

Pohlheim, F. (1971b). Untersuchungen zur Sproßvariation der Cupressaceae. 2. Ploidiechimären an der Haploiden *Thuja gigantea gracilis* nach spontaner Diploidisierung. *Flora.*, **160**, 294–316.

Pohlheim, F. (1971c). Untersuchungen zur Sproßvariation der Cupressaceae. 3.

Quantitative Auswertung des Scheckungsmusters immerspaltender Periklinalchimären. *Flora*, **160**, 360–72.

Pohlheim, F. (1971d). Untersuchungen zur Sproßvariation der Cupressaceae. 5. Das Scheckungverhalten zweier unterschiedlicher Klone von *Juniperus chinensis variegata*. *Wiss. Z. Päd. Hochsch. Potsdam*, **15**, 57–64.

Pohlheim, F. (1971e). Untersuchungen zur Sproßvariation der Cupressaceae. 6. Zum Wechsel von Nadel- und Schuppenform des Blattes. *Wiss. Z. Päd. Hochsch. Potsdam*, **15**, 65–75.

Pohlheim, F. (1971f). *Spiraea bumalda* 'Anthony Waterer' und *Mentha arvensis* 'Variegata' – zwei immerspaltende Periklinalchimären unter den Angiospermen. *Biol. Zbl.*, **90**, 295–319.

Pohlheim, F. (1972a). Untersuchungen zur Sproßvariation der Cupressaceae. 4. Zur Auslese von Mutations-Chimären und Mutanten an der Haploiden *Thuja gigantea gracilis*. *Arch. Züchtungsforsch.*, **2**, 223–35.

Pohlheim, F. (1972b). Überlebensrate und Sproßvariation durch Mutation nach Röntgenbestrahlung haploider und diploider Pflanzen von *Thuja plicata*. *Biol. Rdsch.*, **10**, 200–1.

Pohlheim, F. (1973a). Untersuchungen zur Sproßvariation der Cupressaceae. 7. Auslösung von Perforationen durch Röntgenbestrahlung an der immerspaltenden Periklinalchimäre *Juniperus sabina* 'Variegata'. *Wiss. Z. Päd. Hochsch. Potsdam*, **17**, 79–86.

Pohlheim, F. (1973b). Untersuchungen zur periklinalchimärischen Konstitution von *Pelargonium zonale* 'Freak of Nature'. *Flora*, **162**, 284–94.

Pohlheim, F. (1974). Nachweis von Mischzellen in variegaten Adventivsprossen von *Saintpaulia*, entstanden nach Behandlung isolierter Blätter mit N-Nitroso-N-Methylharnstoff. *Biol. Zbl.*, **93**, 141–8.

Pohlheim, F. (1976). Zur Morphologie des Gynoeceums der Pfropfchimäre *Camellia* + 'Daisy Eagleson'. *Wiss. Z. Päd. Hochsch. Potsdam*, **20**, 57–61.

Pohlheim, F. (1977a). Auslese einer neuen Wuchsform an der Haploiden *Thuja plicata* 'Gracilis'. *Arch. Züchtungsforsch.*, **7**, 311–13.

Pohlheim, F. (1977b). Umlagerungen an der Trichimäre *Pelargonium zonale* 'Freak of Nature' – ein Beitrag zur Herstellung von Plastommutanten. *Wiss. Z. Päd. Hochsch. Potsdam*, **21**, 115–27.

Pohlheim, F. (1977c). Untersuchungen zur Sproßvariation der Cupressaceae. 8. Sind immerspaltende Periklinalchimären bei der haploiden *Thuja plicata* 'Gracilis' möglich? *Flora*, **166**, 177–86.

Pohlheim, F. (1978a). Untersuchungen an Antirrhinum majus wettsteinii, einer an vegetativ haploiden Pflanzen entstanden Sproßvariante. *Zbl. Biol.*, **97**, 53–67.

Pohlheim, F. (1978b). Beobachtungen an Sproßvarianten von *Fragaria vesca* und *Laburnum anagyroides*. *Wiss. Z. Päd. Hochsch. Potsdam*, **22**, 41–5.

Pohlheim, F. (1979). Morphologische, histogenetische, zytologische und genetische Untersuchungen zur Sproßvariation bei höheren Pflanzen. *Biol. Rdsch.*, **17**, 318–20.

Pohlheim, F. (1980a). Zur Sproßvariation bei den Cupressaceae. *Wiss. Z. Humboldt-Universität Berlin, Math.-Nat. R.*, **29**, 295–306.

Pohlheim, F. (1980b). Periklinalchimärische Anthozyanmuster bei *Saintpaulia ionantha* H. Wendl nach NMH-Behandling. *Arch. Züchtungsforsch.*, **10**, 261–9.

Pohlheim, F. (1981a). Genetischer Nachweis einer NMH-induzierten Plastommutation bei Saintpaulia ionantha H. Wendl. *Biol. Rdsch.*, **19**, 47–50.

Pohlheim, F. (1981b). Möglichkeiten der Ausnutzung Haploider in der Pflanzenzüchtung. *Wiss. Z. Päd. Hochsch. Potsdam*, **25**, 85–99.

Pohlheim, F. (1982a). Klonvariabilität durch Chimärenumlagerung und Mutation bei *Dracaena deremensis* (N.E.BR.) Engl. *Arch. Züchtungsforsch.*, **12**, 399–409.

Pohlheim, F. (1982b). Zur Interpretation der Panaschüre von *Sambucus nigra* 'Albo-variegata'. *Folia dendrologica*, **9**, 83–8.

Pohlheim, F. (1983). Vergleichende Untersuchungen zur Änderung der Richtung von Zellteilungen in Blattepidermen. *Biol. Zbl.*, **102**, 323-36.

Pohlheim, F. and Beger, B. (1974). Erhöhung der Mutationsrate im Plastom bei *Saintpaulia* durch N-Nitroso-N-Methylharnstoff. *Biol. Rdsch.*, **12**, 204–6.

Pohlheim, F. and Pohlheim, E. (1974). Über ein Vergilbungsmuster der Blätter von + *Crataegomespilus dardari* und einer Sproßvariante von *Pelargonium zonale* 'Kleiner Liebling'. *Wiss. Z. Päd. Hochsch. Potsdam*, **18**, 47–55.

Pohlheim, F. and Pohlheim, E. (1976). Herstellung von Plastommutanten bei *Saintpaulia ionantha* H. Wendl. *Biochem. Physiol. Pflanzen*, **169**, 377–83.

Pohlheim, F., Pohlheim, E. and Günther, G. (1972). Die Haploide *Pelargonium zonale* 'Kleiner Liebling' als Testsystem für Mutagene. *Wiss. Z. Päd. Hochsch. Potsdam*, **16**, 65–70.

Poiteau, A. (1830). *Cytisus adami*. *Ann. Soc. Hort. Paris.*, **7**, 95–6.

Pötsch, J. (1966a). Über die Auslösung extramutativer Strahlungseffekte an Klonsorten von *Euphorbia pulcherrima* Willd. *Züchter*, **36**, 12–25.

Pötsch, J. (1966b). Das Verhalten von *Abutilon hybridum* hort. 'Andenken an Bonn' nach einmaliger und fraktionierter Röntgenbestrahlung. *Z. Pflanzenzücht.*, **55**, 183–200.

Pötsch, J. (1969). Die Abhängigkeit röntgeninduzierter Histogenese-anomalien von der Höhe Bestrahlungsdosis bei *Pelargonium zonale* Ait. 'Madame Salleron'. *Wiss. Z. Päd. Hochsch. Potsdam*, **13**, 129–37.

Pratt, C. (1960). Changes in structure of a periclinal chromosomal chimera of apple following X-irradiation. *Nature*, **186**, 255–6.

Pratt, C. (1963). Radiation damage and recovery in diploid and cytochimeral varieties of apples. *Radiat. Bot.*, **3**, 193-206.

Pratt, C. (1972). Periderm development and radiation stability of russet-fruited sports of apple. *Hort. Res.*, **12**, 5–12.

Pratt, C., Einset, J. and Gilmer, R. M. (1972a). Reversion in sports of 'Rhode Island Greening' apple. *J. Amer. Soc. Hort. Sci.*, **97**, 292–7.

Pratt, C., Ourecky, D. K. and Einset, J. (1967). Variation in apple cytochimeras. *Amer. J. Bot.*, **54**, 1295–301.

Pratt, C., Way, R. D. and Einset, J. (1975). Chimeral structure of red sports of 'Northern Spy' apple. *J. Amer. Soc. Hort. Sci.*, **100**, 419–22.

Pratt, C., Way, R. D. and Ourecky, D. K. (1972b). Irradiation of colour sports of 'Delicious' and 'Rome' apples. *J. Amer. Soc. Hort. Sci.*, **97**, 268–72.

Preil, W. and Engelhardt, M. (1982). *In vitro*-Entmischung von Chimärenstrukturen durch Suspension-skulturen bei *Euphorbia pulcherrima* Willd. *Gartenbauwissenschaft*, **47**, 241–4.

Prevost, M. jr. (1830). Note sur un Cytise nouveau. *Ann. Soc. Hort. Paris*, **7**, 93–5.

Pusey, J. G. (1962). Sedum 'Aurora' . . a possible chimera. *Nat. Cact. and Succ. J.*, **17**, 11–12.

Renner, O. (1936a). Zur Kenntnis der nichtmendelnden Buntheit der Laubblätter. *Flora.*, **130**, 218–90.

Renner, O. (1936b). Zur Entwicklungsgeschichte randpanaschierter und reingrüne Blätter von *Sambucus, Veronica, Pelargonium, Spiraea, Chlorophytum. Flora*, **130**, 454–66.

Renner, O. and Mitarbeiter (1952). Notizen aus dem Botanischen Garten München-Nymphenburg. *Ber. dtsch. bot. Ges.*, **65**, 296–304.

Renner, O. and Voss, M. (1942). Zur Entwicklungsgeschichte randpanaschierter Formen von *Prunus, Pelargonium, Veronica, Dracaena. Flora.*, **135**, 356–76.

Rieger, R., Michaelis, A, and Green, M. M. (1968). *A glossary of genetics and cytogenetics: classical and molecular.* 3rd edition. George Allen and Unwin, London.

Riemań, G. H. and Darling, H. M. (1947). Trials show how newer potato varieties rate. *Wisconsin Agric. Exp. Sta. Bull.*, **472**, 2–4.

Rieman, G. H., Darling, H. M., Hougas, R. W. and Rominski, M. (1951). Clonal variations in the Chippewa potato variety. *Amer. Potato J.*, **28**, 625–31.

Rischkow, V. L. (1927). Die Verbreitung der Chlorophylls und der Peroxydasegehalt der Epidermis buntblättriger Pflanzen. *Biol. Zbl.*, **47**, 501–12.

Rischkow, V. L. (1936). Buntblättrigen Chimären und der Ursprung des Mesophylls bei Dicotyledonen. *Genetica*, **18**, 313–36.

Robinson, T. R. (1927). An orange chimera. *J. Hered.*, **18**, 48.

Roest, S. and Bokelmann, G. S. (1976). Vegetative propagation of *Solanum tuberosum* L. *in vitro. Potato Res.*, **19**, 173–8.

Roest, S. and Bokelmann, G. S. (1980). *In vitro* adventitious bud techniques for vegetative propagation and mutation breeding of potato (*Solanum tuberosum* L.). I. Vegetative propagation *in vitro* through adventitious shoot formation. *Potato Res.*, **23**, 167–81.

Roselli, G. (1972). Spontaneous mutations and chimeral plants in the olive. *Genetica Agraria*, **26**, 62–74.

Rösler, P. (1928). Histologische Studien am Vegetationspunkt von *Triticum vulgare. Planta*, **5**, 28–69.

Rudloff, C. F. (1931). Pfropfbastarde. *Züchter*, **3**, 15–28.

Sabnis, T. S. (1923). Inheritance of variegation. *Z. Vererbungsl.* **32**, 61–9.

Sabnis, T. S. (1932). Inheritance of variegation. II. *Z. Vererbungsl.*, **62**, 213–31.

Sagawa, Y. and Mehlquist, G. A. L. (1957). The mechanism responsible for some X-ray induced changes in flower colour of the carnation *Dianthus caryophyllus. Amer. J. Bot.*, **44**, 397–403.

Sahli, G. (1916). Die Empfänglichkeit von Pomaceenbastarden, Chimären und intermediaren Formen für *Gymnosporangium. Zbl. Bakt. Abt.*, **2**, **45**, 264–301.

Sarastano, L. and Parrazzani, A. (1911). Di taluni ibrido naturali degli agrumi. *Ann. R. Staz. Sper. Agrumi Frutt. cireale*, **1**, 37–63.

Satina, S. (1944). Periclinal chimeras in *Datura* in relation to development and structure (A) of the style and stigma (B) of calyx and corolla. *Amer. J. Bot.*, **31**, 493–502.

Satina, S. (1945). Periclinal chimeras in *Datura* in relation to the development and structure of the ovule. *Amer. J. Bot.*, **32**, 72–81.

Satina, S. and Blakeslee, A. F. (1941). Periclinal chimeras in *Datura stramonium* in relation to development of leaf and flower. *Amer. J. Bot.*, **28**, 862–71.

Satina, S. and Blakeslee, A. F. (1943). Periclinal chimeras in *Datura* in relation to the development of the carpel. *Amer. J. Bot.*, **30**, 453–62.

Satina, S., Blakeslee, A. F. and Avery, A. G. (1940). Demonstrations of the three germ layers in the shoot apex of *Datura* by means of induced polyploidy in periclinal chimeras. *Amer. J. Bot.*, **27**, 895–905.

Schmidt, A. (1924). Histologische Studien an phanerogamen Vegetationspunkten. *Bot. Arch.*, **8**, 345–404.

Schmidt, M. (1942). Ein Fall genhäufter Chimärenbildung beim Apfel. *Züchter*, **14**, 112–17.

Schwarz, W. (1927). Die Entwicklung des Blattes bei *Plectranthus fruticosus* and *Ligustrum vulgare* und die Theorie der Periklinalchimären. *Planta*, **3**, 499–526.

Sears, B. B. (1980). Elimination of plastids during spermatogenesis and fertilization in the Plant Kingdom. *Plasmid*, **4**, 233–55.

Seeliger, R. (1926). Die Weissdornmispel von Anzig. *Ber. dtsch. bot. Ges.*, **44**, 506–16.

Seeni, S. and Gnanam, A. (1976). Regeneration of green and chimeral shoots from the same explant of heterozygous tobacco grown *in vitro*. *J. Madurai Univ.*, **5**, 43–5.

Seeni, S. and Gnanam, A. (1981). *In vitro* regeneration of chlorophyll chimeras in tomato *Lycopersicon esculentum*. *Can. J. Bot.*, **59**, 1941–3.

Semenuik, P. and Arisumi, T. (1968). Colchicine-induced tetraploid and cytochimeral roses. *Bot. Gaz.*, **129**, 190–3.

Shamel, A. D. (1932). A pink fruited lemon. *J. Hered.* **23**, 23–7.

Shamel, A. D., Pomeroy, C. S. and Caryl, R. E. (1926). Bud selection in the Washington navel orange. VI. Progeny test of a dual limb variation. *J. Hered.*, **17**, 59–65.

Sharma, A. K. and Sharma, A. (1972). *Chromosome techniques: theory and practice*. 2nd edition. Butterworths, London.

Siddiq, E. A. (1967). Colchicine induced autotetraploids in *Sorghum vulgare*. *Indian J. Genet. Plant Breed.*, **27**, 442–52.

Simmonds, N. W. (1965). Chimeral potato mutants. *J. Hered.*, **56**, 139–42.

Simmonds, N. W. (1969). Variegated mutant plastid chimeras of potatoes. *Heredity*, **24**, 303–6.

Skene, K. G. M. and Barlass, M. (1983). Studies on the fragmented shoot apices of grapevine. IV. Separation of phenotypes in a periclinal chimera *in vitro*. *J. exp. Bot.*, **34**, 1271–80.

Smith, R. H. and Norris, R. E. (1983). *In vitro* propagation of African violet chimeras. *Hort Science*, **18**, 436–7.

Sneath, P. H. A. (1968). Numerical taxonomic study of the graft chimera + *Laburnocytisus adamii*. *Proc. Linn. Soc. London*, **179**, 83–96.

Spiegel-Roy, P. (1979). On the chimeral nature of the Shamouti orange. *Euphytica*, **28**, 361–6.

Sree Ramulu, K., Derreux, M., Ancora, G. and Laneri, U. (1976a). Chimerism in *Lycopersicon peruvianum* plants regenerated from *in vitro* cultures of anthers and stem internodes. *Z. Pflanzenzücht.*, **76**, 299–319.

Sree Ramulu, K., Derreux, M. and de Martinis, P. (1976b). Origin and genetic analysis of plants regenerated *in vitro* from periclinal chimeras of *Lycopersicon peruvianum*. *Z. Pflanzenzücht.*, **77**, 112–20.

Stahl, J. L. (1946). Chimeras of Bowen blackberry. *J. Hered.*, **37**, 51–3.

Staudt, G. (1973). Ploidiechimären bei europaischen *Vitis vinifera* Sorten. *Vitis*, **12**, 89–92.

Stebbins, G. L. (1971). *Chromosomal evolution in higher plants*. Edward Arnold, London.

Steffensen, D. M. (1968). A reconstruction of cell development in the shoot apex of maize. *Amer. J. Bot.*, **55**, 354–69.

Steinberg, C. (1950). Richerche sull 'istogenesi dell 'apice vegetativo di alcune specie del genere *Solanum*. *Nuovo G. bot. ital.* (N.S.) **57**, 319–34.

Stewart, R. N (1965). The origin and transmission of a series of plastogene mutants in *Dianthus* and *Euphorbia*. *Genetics*, **52**, 925–47.

Stewart, R. N. and Arisumi, T. (1966). Genetic and histogenic determination of pink bract colour in *Poinsettia*. *J. Hered.*, **57**, 217–220.

Stewart, R. N. and Burk, L. G. (1970). Independence of tissues derived from apical layers in ontogeny of the tobacco leaf and ovary. *Amer. J. Bot.*, **57**, 1010–16.

Stewart, R. N. and Dermen, H. (1970a). Determination of number and mitotic activity of shoot apical initial cells by analysis of mericlinal chimeras. *Amer. J. Bot.*, **57**, 816–26.

Stewart, R. N. and Dermen, H. (1970b). Somatic genetic analysis of the apical layers of chimeral sports in *Chrysanthemum* by experimental production of adventitious shoots. *Amer. J. Bot.*, **57**, 1061–71.

Stewart, R. N. and Dermen, H. (1975). Flexibility in ontogeny as shown by the contribution of the shoot apical layers to leaves of periclinal chimeras. *Amer. J. Bot.*, **62**, 935–47.

Stewart, R. N. and Dermen, H. (1979). Ontogeny in monocotyledons as revealed by studies of the developmental anatomy of periclinal chloroplast chimeras. *Amer. J. Bot.*, **66**, 47–58.

Stewart, R. N., Meyer, F. G. and Dermen, H. (1972). Camellia + 'Daisy Eagleson', a graft chimera of *Camellia sasanqua* and *C. japonica. Amer. J. Bot.*, **59**, 515–24.

Stewart, R. N., Semenuik, P. and Dermen, H. (1974). Competition and accommodation between apical layers and their derivatives in the ontogeny of chimeral shoots of *Pelargonium* × *hortorum. Amer. J. Bot.*, **61**, 54-67.

Strasburger, E. (1906). Histologische Beiträge zur Vererbungsfrage. I. Typische und allotypische Kernteilung. *Jb. wiss. Bot.*, **42**, 1–71.

Strasburger, E. (1907). Über die Individualität der Chromosomen und die Pfropfhybridenfrage. *Jb. wiss. Bot.*, **44**, 482–556.

Strasburger, E. (1909). Meine Stellungnahme zur Frage der Pfropfbastarde. *Ber. dtsch. bot. Ges.*, **27**, 511–28.

Straub, J. (1940). Die Auslösung von polyploidem *Pisum sativum. Ber. dtsch. bot. Ges.*, **58**, 430–6.

Sussex, I. M. (1955). Morphogenesis in *Solanum tuberosum* L. : Apical structure and developmental pattern of the juvenile shoot. *Phytomorphology*, **5**, 253–73.

Swingle, C. F. (1927). Graft hybrids in plants. *J. Hered.*, **18**, 73–94.

Tanaka, T. (1927a). Bizzaria – a clear case of periclinal chimera. *J. Genet.*, **18**, 77–85.

Tanaka, T. (1927b). Bud variation, chimera, somatic segregation, and chromosomal mutation in fruit trees. *Agric. and Hortic.*, **2**, 3–10. (Japanese).

Thielke, C. (1948). Beiträge zur Entwicklungsgeschichte und zur Physiologie panaschierter Blätter. *Planta.* **36**, 2–33.

Thielke, C. (1951). Über die Möglichkeiten der Periklinalchimärenbildung bei Gräsern. *Planta*, **39**, 402–30.

Thielke, C. (1954). Die histologische Struktur des Sproßvegetationskegel einiger Commelinaceen unter Berücksichtigung panaschierter Formen. *Planta*, **54**, 18–74.

Thompson, M. M. and Olmo, H. P. (1963). Cytohistological studies of cytochimeric and tetraploid grapes. *Amer. J. Bot.*, **50**, 901–6.

Tilney-Bassett, R. A. E. (1963a). The structure of periclinal chimeras *Heredity*, **18**, 265–85.

Tilney-Bassett, R. A. E. (1963b). Genetics and plastid physiology in *Pelargonium. Heredity*, **18**, 485–504.

Tilney-Bassett, R. A. E. (1984). The genetic evidence for nuclear control of chloroplast biogenesis in higher plants. In: *Chloroplast biogenesis* (Ed. R. J. Ellis), 13–50. Cambridge University Press, Cambridge.

Tischler, G. (1903). Über eine merkwürdige Wachstumerscheinung in den Samenanlagen von *Cytisus adami* Poit. *Ber. dtsch. bot. Ges.*, **21**, 82–9.

Trelease, W. (1908). Variegation in the Agaveae. *Wiesner-Festschrift, Wien*, 332–56.

Tsanev, T. (1976). Peach – a chimera mutant. *Priroda, Bulgaria*, **25**, 65–6. (Bulgarian)

Ufer, M. (1936). Erblichkeitsuntersuchungen an 'Freak of Nature'. Ein Beitrag zur Frage der nichtmendelnden Vererbung chlorophyll-defekter Formen von *Pelargonium. Z. Vererbungsl.*, **71**, 281–98.

Van Harten, A. M. (1972). A suggested method for investigating L I constitution in periclinal potato chimeras. *Potato Res.*, **15**, 73–5.

Van Harten, A. M. (1978). *Mutation breeding techniques and behaviour of irradiated shoot apices of potato.* Pudoc, Wageningen.

Van Harten, A. M. and Bouter, H. (1973). Dihaploid potatoes in mutation breeding: some preliminary results. *Euphytica*, **22**, 1–7.

Van Harten, A. M., Bouter, H, and Broertjes, C. (1981). *In vitro* adventitious bud techniques for vegetative propagation and mutation breeding of potato (*Solanum tuberosum* L.). II. Significance for mutation breeding. *Euphytica*, **30**, 1–8.

Van Harten, A. M., Bouter, H. and Schut, B. (1973). Ivy leaf of potato (*Solanum tuberosum*), a radiation-induced dominant mutation for leaf shape. *Radiat. Bot.*, **13**, 287–92.

Van Harten, A. M., Bouter, H. and van Ommeren, A. (1972). Preventing chimerism in potato (*Solanum tuberosum* L.). *Euphytica*, **21**, 11–21.

Vasil'ev, A. E. (1965). Induced inversion of components in poplar chimeras. *Dokl. Akad. Nauk. SSSR.*, **164**, 1401–4. (Russian)

Verkerk, K. (1971). Chimerism of the tomato plant after seed irradiation with fast neutrons. *Neth. J. Agric. Sci.*, **19**, 197–203.

Walkof, C. (1964). A pericarp chimera in the tomato. *Can. J. Genet. and Cytol.*, **6**, 46–51.

Wanjari, K. B. and Phadnis, B. A. (1973). Colchicine induced cytochimera in brinjal (*Solanum melongea* L.). *PKV Res. J.*, **1**. 136–7.

Weber, W., Darling, H. and Rieman, G. (1947). Russet Sebago potato developed. *Wisconsin Agric. Exp. Sta. Bull.*, **472**, 1–2.

Webster, W. H. and Rieman, G. H. (1949). Unusual variegation in the Sebago potato. I. Somatic mutations. *Amer. Potato J.*, **26**, 104.

Weiss, F. E. (1925). On the leaf-tissues of the graft hybrids of *Crataegomespilus asnieresii* and *Crataego-mespilus dardari*. *Mem. proc. Manch. Lit. and Philos. Soc.*, **69**, 73–8.

Weiss, F. E. (1930). The problem of graft hybrids and chimeras. *Biol. Rev. Camb. Phil, Soc.*, **5**, 231–71.

Wenninger, M. J. (1971). *Polyhedron models.* Cambridge University Press, London.

Winkler, H. (1907). Über Pfropfbastarde und pflanzliche Chimären. *Ber. dtsch. bot. Ges.*, **25**, 568–76.

Winkler, H. (1908). Solanum tubingense, ein echter Pfropfbastarde zwischen Tomate und Nachtschatten. *Ber. dtsch. bot. Ges.*, **26a**, 595–608.

Winkler, H. (1909). Weitere Mitteilungen über Pfropfbastarde. *Z. Bot.*, **1**, 315–45.

Winkler, H. (1910a). Über das Wesen der Pfropfbastarde. *Ber. dtsch. bot. Ges.*, **28**, 116–18.

Winkler, H. (1910b). Über die Nachkommenschaft der Solanum-Pfropfbastarde und die Chromosomen-Zahlen ihrer Keimzellen. *Z. Bot.*, **2**, 1–38.

Winkler, H. (1916). Über die experimentelle Erzeugung von Pflanzen mit abweichenden Chromosomenzahlen. *Z. Bot.*, **8**, 417.

Winkler, H. (1934). Über zwei Solanum-Chimären mit Burdonen-epidermis. *Planta*, **21**, 613–56.

Winkler, H. (1935). Chimären und Burdonen. Die Lösung des Pfropfbastardproblems. *Biologe*, **9**, 279–90.

Winkler, H. (1938). Über einen Burdonen von *Solanum lycopersicum* und *Solanum nigrum*. *Planta*, **27**, 680–707.

Wolff, C. F. (1759). *Theoria generationis.* Wilhelm Engelman, Leipzig.

Yamane, H., Kurihara, A. and Tanaka, R. (1978). Studies on polyploidy breeding in grapes. Part 1. Chromosome numbers of large berried grape varieties grown in Japan. *Bull Fruit Tree Res. Stn. (Minist. Agric. For.)* Ser. E, 1–8.

178   References

Yamashita, K. (1979). The chimeral nature of the *Citrus* cv. Kobayashi Mikan with special reference to its epidermal structure as observed by scanning electron microscopy. *J. Jpn. Soc. Hort. Sci.*, **48**, 169–78. (Japanese)

Yamashita, K. (1983). Chimerism of Kobayashi-mikan (*Citrus natsudaidai* × *unshiu*) judged from isozyme patterns in organs and tissues. *J. Jpn. Soc. Hort. Sci.*, **52**, 223–30.

Yasui, K. (1929). Studies on the maternal influence of plastid characters in *Hosta japonica albomarginata* and its derivatives. *Cytologia, Tokyo*, **1**, 192–215.

Yeager, A. F. and Meader, E. M. (1956). A flesh-colour chimera in the peach. *J. Hered.*, **47**, 77–8.

Zimmerman, P. W. and Hitchcock. A. E. (1951). Rose 'sports' from adventitious buds. *Contr. Boyce Thompson Inst.*, **16**, 221–4.

Zobel, D. B. (1974). A periderm-color chimera in *Abies*. *Can. J. Bot.*, **52**, 1435–7.

Indices

Subject Index

Index to Taxa

Insects

Macrosiphum euphorbiae (Potato aphid) 17
Trialeuroides vaporariorum (White fly) 17

Fungi

Cladosporium fulvum 14
Gymnosporangium confusum 14, 174
Podosphaera oxycanthae 14
Septoria lycopersicii 14

Viruses

Pelargonium net vein agent 57
Pelargonium petal streak agent 57

Gymnosperms

Abies concolor (White fur) 31, 178
Araucariaceae 110

Chamaecyparis lawsoniana (Lawson cypress)
 'Argenteo-variegata' 109
Chamaecyparis nootkatensis (Nootka or Stinking cypress)
 'Argenteo-variegata' 109
Chamaecyparis pisifera (Sawara cypress)
 'Albo-variegata' 109
 'Plumosa argentea' 109
Chamaecyparis thyoides (White cypress)
 'Variegata' 109
Cupressaceae 108, 171-2
Cycas revoluta (Sago palm) 19

Ephedraceae 110

Ginkgo biloba (Maidenhair tree) 164
Gnetaceae 110

Juniperus chinensis (Chinese juniper)
 'Plumosa argenteo-variegata' 109
 'Plumosa aureo-variegata' 109
 'Variegata' 109
Juniperus horizontalis (Creeping juniper 'Variegata' 109
Juniperus sabina (Savin)
 'Variegata' 76, 109-10, 172
Juniperus virginiana (Pencil or Eastern red cedar)
 'Triomphe d'Angers' 109

Thuja plicata (Western red cedar)
 'Excelsa' 74, 171
 'Gracilis' 32, 74-6, 171-2

Monocotyledons

Acorus gramineus (Grassy-leaved sweet flag)
 'Variegatus' 107
Agave americana (Century plant)
 'Mediovariegata' 95
Aglaonema modestum (Chinese evergreen)
 'Variegata' 107
Ananas comosus (sativus) (Pineapple) 66, 68, 70, 84, 95
 'Cayenne' 64, 167
Arrhenatherum elatius (Oat grass)
 'Variegatum' 84, 88, 95
Arundo donax (Reed grass)
 'Variegata' 84
Apidistra lurida (elatior) (Cast-iron plant or Parlour palm)
 'Variegata' 107
Avena sativa (Oat) 88

Bambusa verticillata (Bamboo) 89

Carex morrowi (Japanese sedge grass) 84, 95
Carex riparia (Great pond sedge)
 'Variegata' 95

Index to Authors